T0213495

Undergraduate Lecture Notes in Physics

Series Editors

Neil Ashby, University of Colorado, Boulder, CO, USA

William Brantley, Department of Physics, Furman University, Greenville, SC, USA

Matthew Deady, Physics Program, Bard College, Annandale-on-Hudson, NY, USA

Michael Fowler, Department of Physics, University of Virginia, Charlottesville, VA, USA

Morten Hjorth-Jensen, Department of Physics, University of Oslo, Oslo, Norway

Michael Inglis, Department of Physical Sciences, SUNY Suffolk County Community College, Selden, NY, USA

Barry Luokkala (iD), Department of Physics, Carnegie Mellon University, Pittsburgh, PA, USA

Undergraduate Lecture Notes in Physics (ULNP) publishes authoritative texts covering topics throughout pure and applied physics. Each title in the series is suitable as a basis for undergraduate instruction, typically containing practice problems, worked examples, chapter summaries, and suggestions for further reading.

ULNP titles must provide at least one of the following:

- An exceptionally clear and concise treatment of a standard undergraduate subject.
- A solid undergraduate-level introduction to a graduate, advanced, or non-standard subject.
- A novel perspective or an unusual approach to teaching a subject.

ULNP especially encourages new, original, and idiosyncratic approaches to physics teaching at the undergraduate level.

The purpose of ULNP is to provide intriguing, absorbing books that will continue to be the reader's preferred reference throughout their academic career.

More information about this series at https://link.springer.com/bookseries/8917

Yury Deshko

Special Relativity

For Inquiring Minds

 Springer

Yury Deshko
Weehawken, NJ, USA

ISSN 2192-4791   ISSN 2192-4805   (electronic)
Undergraduate Lecture Notes in Physics
ISBN 978-3-030-91141-6   ISBN 978-3-030-91142-3   (eBook)
https://doi.org/10.1007/978-3-030-91142-3

© The Editor(s) (if applicable) and The Author(s), under exclusive license to Springer Nature Switzerland
AG 2022
This work is subject to copyright. All rights are solely and exclusively licensed by the Publisher, whether
the whole or part of the material is concerned, specifically the rights of translation, reprinting, reuse
of illustrations, recitation, broadcasting, reproduction on microfilms or in any other physical way, and
transmission or information storage and retrieval, electronic adaptation, computer software, or by similar
or dissimilar methodology now known or hereafter developed.
The use of general descriptive names, registered names, trademarks, service marks, etc. in this publication
does not imply, even in the absence of a specific statement, that such names are exempt from the relevant
protective laws and regulations and therefore free for general use.
The publisher, the authors and the editors are safe to assume that the advice and information in this book
are believed to be true and accurate at the date of publication. Neither the publisher nor the authors or
the editors give a warranty, expressed or implied, with respect to the material contained herein or for any
errors or omissions that may have been made. The publisher remains neutral with regard to jurisdictional
claims in published maps and institutional affiliations.

This Springer imprint is published by the registered company Springer Nature Switzerland AG
The registered company address is: Gewerbestrasse 11, 6330 Cham, Switzerland

Dedicated to Sir Hermann Bondi
1919–2005

Preface

The material presented in this book has sprung from the soil of teaching, and therein lies its strength. For a number of years, I had both pleasure and privilege of teaching an introduction into the special theory of relativity in summer schools of Johns Hopkins University (2013–2015) and Columbia University (2019–2021). The audience consisted of motivated students with the knowledge of high school algebra, trigonometry, and Galileo-Newtonian mechanics. The course ran, in total, for about 60 hours.

The choice of a teaching approach was significantly influenced by the works of Sir Hermann Bondi.[1] Bondi's book, *Relativity and Common Sense*, was used as the foundation upon which the course notes were built and expanded. In a certain sense, this textbook is a homage to the contribution of Sir Hermann Bondi into the field of teaching the special theory of relativity.

The background knowledge and the way mind works is different for different people. Every reader requires an individualized approach; however, there are many more readers than there are books on a given subject. The best way to alleviate this problem, in my view, is to encourage every reader to look through many books, to read explanations of a subject from different angles. The special theory of relativity has a very rich—and ever growing—selection of books. If the way this book presents the special theory of relativity resonates with some readers, its mission will be accomplished and its existence will be justified.

Readers must have no illusions: special relativity is not an easy subject. After all, there are reasons why some physicists in the early twentieth century struggled to accept it, and today, not every physics major "gets it" after taking a course in college.

There is no "royal road" that would make the study of relativity easy or throughout fun. A student has to put in hard work, otherwise only a deceptive feeling of understanding will remain. To help with better mastery of the subject, the book offers numerous examples, illustrations, and problems with solutions.

---

[1] H. Bondi *The teaching of special relativity* 1966 Phys. Educ. 1 223.

Fig. 1 Two possible paths of reading the book. The second path is recommended when students need a quicker introduction into the special theory of relativity

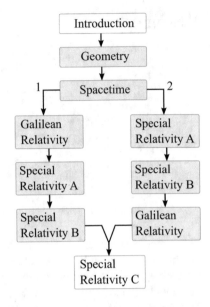

As illustrated in Fig. 1, there are two paths that can be followed when reading the book. The first path takes students through Galilean relativity, before discussing special relativity; it is a normal, recommended order. Its downside is the delay before the discussion of the special theory of relativity begins. Such a delay may play a role for some students, eager to dive into the exciting world of Einstein's special theory of relativity.

The second path takes students to the special theory of relativity quicker, perhaps at the expense of some continuity in logic. It is a valid path and works well, provided the Galilean relativity is not omitted. In fact, discussing Galilean relativity after getting acquainted with the special theory of relativity allows students to better see how the former is contained in the latter as a special case.

Finally, if a short introduction into the special theory of relativity is required, one can skip the *Introduction* and the last chapter—"Special Relativity C." The sections with an asterisk (e.g., Compton Scattering*) can also be omitted in short introduction.

I know for a fact that there are many people who, like my former students, are *curious*, *bright*, and *studious*. If you are one of them, and you are interested in exploring relativity, you may find this book helpful.

Weehawken, NJ, USA
August 2021

Yury Deshko

Acknowledgment

Firstly, I want to thank all my students—they inspired and motivated me to write this book. Secondly, I am indebted to the first readers of the manuscript: Dr. Alexander Punnoose, Dr. Mike D. Schneider, Dr. Alex Rylyakov, Dr. Mikhail Makouski, Dr. Maxim Molchan, and Mr. Simai Jia. Finally, my gratitude goes to my wife, Anna, who supported my teaching and writing efforts all these years.

Contents

Part I
Relativity

Chapter 1
Introduction

> *We are at the very beginning of time for the human race. It is not unreasonable that we grapple with problems. But there are tens of thousands of years in the future. Our responsibility is to do what we can, learn what we can, improve the solutions, and pass them on.*
>
> R. Feynman, *"The Value of Science"*

Abstract This chapter gives a birds-eye view of the special theory of relativity. Each section presents, in a condensed form, some aspect of the subject: from historical to physical and mathematical. The goal of these sections is not to be exhaustive, but to stimulate further reading on each topic, either in this book or elsewhere. The creation of the special theory of relativity was an epochal event in the history of human thought. Anyone who devotes time to the study of its physical, mathematical, or historical aspects will be rewarded manifold.

1.1 Brief Historical Context

The special theory of relativity was created at the dawn of the twentieth century. In 1905 Albert Einstein published a paper containing the analysis of the most fundamental physical concepts: the measurements of time intervals and lengths. He showed how common sense view of simultaneity is wrong, and how the usual methods of measuring times and lengths are affected by motion. The ideas about space and time, that seemed evident in the times of Galileo and Newton, were found to be not true.

Albert Einstein was not alone in the quest for a deeper understanding of time and space. In the heart of the mathematical aspect of the special theory of relativity lie equations called *Lorentz transformation*, named after Dutch theoretical physicist Antoon Hendrik Lorentz, who derived them in 1895. Henri Poincare, a French

© The Author(s), under exclusive license to Springer Nature Switzerland AG 2022
Y. Deshko, *Special Relativity*, Undergraduate Lecture Notes in Physics,
https://doi.org/10.1007/978-3-030-91142-3_1

mathematician,[1] also contributed to the analysis of Lorentz transformation, to the problem of time measurements, and to the meaning of simultaneity. As early as 1904 Poincare advocated the use of the principle of relativity as the guiding principle of scientific inquiry.

The time period when Lorentz, Poincare, and Einstein were actively working on the issues of light, time, and relativity was rich with new scientific discoveries and technological advances. Electron, the first subatomic particle, was discovered by J. J. Thomson in 1897—just a year after the discovery of radioactivity by Henri Becquerel. The radioactive sources of fast electrons gave physicists an opportunity to experimentally study the properties of massive and electrically charged particles moving at the speeds comparable to the speed of light. Several early theories explained the origin of the electron's mass and showed how the latter could depend on the electron's speed. The experiments performed by Walter Kaufmann and Alfred Bucherer in 1901–1908 helped decide which electron model was correct, and gave the first experimental support to the special theory of relativity.[2]

In 1880–1890, during Albert Einstein's early childhood, Heinrich Hertz experimentally confirmed the existence of electromagnetic waves, and Guglielmo Marconi began applying the newly discovered "Hertzian waves" for wireless telegraphy. The wireless technology rapidly developed, and, by the time Einstein got his first job as a technical expert in Swiss patent office in 1902, found applications in such important areas as marine navigation and transatlantic communication. Wired and wireless electromagnetic signals were used to ensure that clocks in different places—city halls, post offices, railroad stations, and ships at sea—were synchronized. Thus, for Albert Einstein the problems of clock synchronization and electromagnetic signaling were well known.

Henri Poincare was also actively involved in the application of telegraphic signals for time distribution. Since 1893 he worked in *Bureau des Longitudes*, French institution responsible for setting up global network of synchronized clocks, and wrote extensively on the issues of time measurement and simultaneity.

Apart from the practical applications of electromagnetic waves, the fundamental problem of the nature of these waves intrigued many scientists, including Lorentz, Poincare, and Einstein. There is a certain similarity between electromagnetic waves and sound waves. Acoustic sound waves require air for their propagation, so it was natural to ask: What kind of medium is required for the propagation of electromagnetic waves? The hypothesized medium was called *aether* or *ether* and was the subject of active theoretical and experimental research. The aether was thought to fill all the space not occupied by atoms of regular matter, from interstellar voids to the inter-atomic spaces. The expected effects of the aether on normal bodies was extremely small and required sensitive measurements.

---

[1] Henri Poincare was incredibly prolific in mathematics, physics, philosophy and engineering.

[2] This is a simplification of actual historical events. See Miller A. *Albert Einstein's Special Theory of Relativity*, Springer-Verlag New York, 1998.

The methods of optical interferometry, still important today, played an important role in many experiments related to the study of the aether effects. This technique was used by Hippolyte Fizeau in 1851 to confirm that moving regular matter (flowing water in his experiment) could "drag" the aether to some degree. Many years later, in 1886, Albert Michelson and Edward Morley reproduced the same results with higher accuracy and in a more careful experiment, again using the optical interferometry. The next year Michelson and Morley performed their most famous experiment, which was expected to detect much stronger effect of the aether. The result puzzled everyone: There was no effect at all, *as if the aether did not exist.*

In 1889 an Irish physicist George Francis FitzGerald speculated that all bodies contracted along the direction of their motion due to the interaction with the aether. The surprising result of Michelson-Morley experiment could then be explained by the contraction of their measurement apparatus due to the motion of earth through the aether. Antoon Lorentz provided theoretical details of the proposed effect in 1895, and this hypothesis of aether-induced length change became known as FitzGerald-Lorentz contraction.[3]

The earlier mentioned 1905 paper of Albert Einstein provided a completely different view of the problems. It showed how to reconcile the results of the experiments of Fizeau and Michelson-Morley. It gave a clear physical meaning of the Lorentz transformation, the idea of simultaneity, the measurement of time duration, the lengths and masses of moving objects. It created the special theory of relativity as we know it today—*Einstein's special theory of relativity.*

As Einstein's special theory of relativity became more widely known, it met both support and resistance. Support came from physicists and mathematicians who understood and appreciated the profound nature of new ideas. The resistance came mostly from people overwhelmed by radically new views on space and time. The resistance eventually died out, and the theory continued to be mathematically developed, experimentally tested, and actively popularized.

An important contribution to the mathematical presentation of the special theory of relativity was made by Hermann Minkowski, who once was Einstein's professor of mathematics. Minkowski, an expert in geometrical methods, showed in 1908 how to represent the special theory of relativity as a variant of *non-Euclidean geometry.* Non-Euclidean geometries were of great interest to the mathematicians of that era. Today geometric approach to the special theory of relativity is standard; it is actively used in this book.

Immediately after its creation, the special theory of relativity became the subject of vigorous discussions among physicists, as well as non-physicists. A large number of popular expositions of the main ideas and results of the theory appeared, including the books by Albert Einstein. New books on the subject are published every year, making the special theory of relativity one of the most popular topics in physics.

---

[3] Sometimes called FitzGerald-Larmor-Lorentz contraction hypothesis, to emphasize the contribution of Joseph Larmor into this problem.

In 1964 Sir Hermann Bondi, a British-Austrian mathematician and cosmologist, introduced a new method for teaching the special theory of relativity, now known as Bondi's k-calculus. His book, *"Relativity and Common Sense"*, became the inspiration for this work.

Practice showed that k-calculus, combined with Minkowski's geometrical approach, provides an effective method of teaching the special theory of relativity even in high-school. A student with a solid knowledge of algebra, trigonometry, and Newtonian mechanics can learn the core of the special theory of relativity. This book serves as the guide on this path (Fig. 1.1).

1.2   Prelude

Two space stations, *Aether* (A) and *Helios* (H),[4] were freely floating in space, observing an unknown object X (see Fig. 1.2). The distance AH between the stations, as well as the distances AX and HX from each station to the object X remained unchanged. Both stations were using identical clocks.

When the clock on the board of *Aether* read $a_1 = 300$ seconds, a radio signal was received from the object X. It was followed by another signal at $a_2 = 527$ seconds, and yet another at $a_3 = 614$ seconds.

The log of events on *Helios* was telling a different story: The first signal was received at $h_1 = 419$ seconds, the second at $h_2 = 646$ seconds, and the third at $h_3 = 733$ seconds.

After each station accounted for the time required for the radio signal from the object X to reach it, new *adjusted* values were obtained, shown in the Table 1.1. It then became clear that the signals received by two different stations were likely caused by the same events.

The differences between the immediate (raw) measurements of the times of the signals' arrivals were due to the difference in the locations of the stations relative to the object X. This factor can be taken into account if the distances from *Aether* to X and from *Helios* to X are known.

1.2.1   Static Observers

The stations *Aether* and *Helios* are *not* identical observers, but they are, in a certain sense, *equivalent*. There is nothing that makes the measurements made by *Aether* more valid than the measurements made by *Helios*, and vice versa. Both observers should arrive at the same picture of events happening on X, having done appropriate corrections to the results of their measurements. The locations of the stations are

---

[4] In Ancient Greek mythology Aether is the deity of upper atmosphere; Helios is the god of Sun.

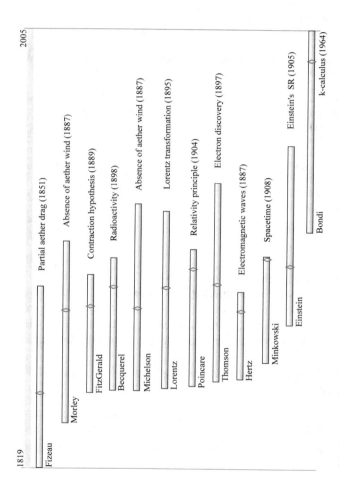

Fig. 1.1 Prominent figures and their important contributions into the development of the special theory of relativity. The unfortunately short life of a German physicist Heinrich Rudolf Hertz is striking

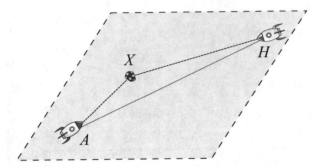

Fig. 1.2 Two space stations, A and H, observing an object X. The distance between each pair of objects remains constant

	Signal	A-time (s)	H-time (s)
Table 1.1 Times of detection of three successive signals from the object X, as measured by two space station. Both "raw" and "adjusted" values are shown. The adjustment takes into account the time required for each signal to reach a space station		Raw	
	1	$a_1 = 300$	$h_1 = 419$
	2	$a_2 = 527$	$h_2 = 646$
	3	$a_3 = 614$	$h_3 = 733$
		Adjusted	
	1	$A_1 = 0$	$H_1 = 0$
	2	$A_2 = 227$	$H_2 = 227$
	3	$A_3 = 314$	$H_3 = 314$

irrelevant. This is also true for *any other observer*, which is at rest relative to *Aether* and *Helios*. Any number of such observers, all in mutual rest, can be imagined. We will call the whole family of such *mutually stationary* observers **static observers**. As an example, if we disregard the rotation of the Earth around its axis, we can consider astrophysical observatories in different geographical locations as static observers.

There is nothing special about the events happening to the object X. The space stations could observe any other event. It can be said that:

▷ **Equivalence of Static Observers**

All static observers are equivalent in their observations/measurements of any event.

The equivalence of static observers is related to the well established experimental fact of the *uniformity of space*. The uniformity of space means that identical experiments (or physical processes) produce identical results (or

(continued)

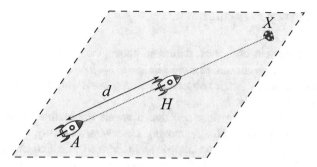

Fig. 1.3 Two space stations, A and H, are observing an unknown object X. All three are lying on the same straight line. The distance to the object measured by the spaceship A is related to the distance measured by the spaceship H: $x_a = x_h + d$

events), regardless of the location where the experiment (or a process) takes place. Identical atoms at different places will emit identical light, identical clocks will run at identical rate, and so on.

The results of the measurements of positions and times of events may differ for different observers. For example, the raw times of detection of the first signal by the stations *Aether* and *Helios* are different: $a_1 \neq h_1$. However, the observers can either infer some results that are equal for both of them, or they can relate their results through equations. In the case of the space stations *Aether* and *Helios*, the observers arrived at the same detection times after the adjustments (e.g., $A_1 = H_1$).

To see how two observers can relate the results of their measurements, consider a special case, illustrated in Fig. 1.3. If the distance between the stations is d and the speed of radio signals is c, then the relationships between the time and position of any event at X, as measured by *Aether* and *Helios*, are given by

$$x_a = x_h + d,$$

$$t_a = t_h + d/c + t_{off}.$$

The subscripts a and h are for the measurements done by *Aether* and *Helios*, respectively. The time t_{off} corresponds to a possible constant offset between the clocks of two space stations. When the clocks are synchronized $t_{off} = 0$.

The conclusion we make is the following: *None of the static observers is privileged. All static observers are equivalent in their measurements of the same events. Everyone observes the same reality, although from a different point of view.*

1.2.2 Relativity Principle

The idea of equivalent observers, described above, can be taken one step further. Different observers are allowed to *move relative to each other with constant velocity*. The mutually stationary (static) observers will represent a special case of "motion" with zero relative velocity.

It has been noticed long time ago that it seems impossible to establish who, among the multitude of mutually moving observers, is "truly in motion". The motion, it must be noted, must be *uniform* and *rectilinear*. Galileo Galilei expressed this idea in 1632, using the example of an observer in a closed room of a large ship which could *"proceed with any speed you like, so long as the motion is uniform and not fluctuating this way and that"*.[5] Galileo pointed out that there could be no experiment or observation performed in the room which would tell whether the ship, and therefore the observer in the room, was moving. Everything should remain the same, regardless of the smooth motion of the ship.

The idea of Galileo became an important guiding principle in Newtonian mechanics. The principle, as applied to mechanics, can be stated as follows:

▷ **Principle of Galilean Relativity**

All laws of mechanics are the same for all observers, regardless of their relative motion with constant velocity.

The way Galileo formulated his view is very general. It is applicable to *all* physical phenomena and *all* speeds, even the speeds beyond the speed of light.

As physics evolved, new phenomena were discovered and the laws of optics, electricity, magnetism, heat and others were determined. Unsuccessful attempts were made to detect "true motion" using experiments beyond mechanics. Finally, the idea of Galileo was explicitly extended to *all* physical laws, not just mechanics. In 1904 Henri Poincare wrote:[6]

---

[5] *Dialogue Concerning the Two Chief World Systems*, translated by Stillman Drake, University of California Press, 1953, pp. 186–187.

[6] H. Poincare, *The Principles of Mathematical Physics*. Translation in *The Foundations of Science (The Value of Science)*, 1913, New York: Science Press, Chap. 7–9, pp. 297–320.

> **Poincare Relativity Principle**

"...the laws of physical phenomena must be the same for a stationary observer as for an observer carried along in a uniform motion of translation; so that we have not and can not have any means of discerning whether or not we are carried along in such a motion."

The laws of electrodynamics, discovered by James Clerk Maxwell around 1862, tell that light propagates like a wave with a certain speed. Experiments indicated that this speed, unlike the speed of objects in mechanics, might be *the same for all observers*.

Using the constancy of the speed of light as a fundamental postulate, Einstein's special theory of relativity provides *the correct way* to relate the measurements of positions and times (and other quantities derived from them, like speed or angular momentum) of the same events, performed by observers in uniform rectilinear relative motion. The *correct way* means *in agreement with wide range of physical experience*, including electrodynamics and high energy physics. Special relativity reveals how old common-sense conceptions of distance and time interval, as well as many other quantities based on them, are only approximately valid and are *not* in agreement with current physical knowledge.

1.3 Quick Tour Through Special Relativity

To get a glimpse of the main results of the special theory of relativity, we will consider a specific scenario, illustrated in Fig. 1.4. Two spaceships are traveling from the solar system to the binary star Sirius. It takes approximately 9 years for the light to cover the distance between these two places. Let's assume that the speed of the first spaceship is $v_1 = 100\,000$ km/s (1/3 of the speed of light), the speed of the second spaceship is $v_2 = 150{,}000$ km/s (1/2 of the speed of light). The spaceships start their journey at the same time; their identical clocks show the same zero time at the launch.

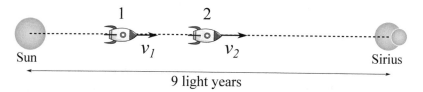

Fig. 1.4 Two spaceships are traveling from the Sun to Sirius with the speeds $v_1 = c/3$ and $v_2 = c/2$

The predictions of the special theory of relativity will be given without derivation or justification; we will simply compare the results of pre-relativistic Galileo-Newtonian physics to the results of the special theory of relativity. Later chapters will provide proper explanations of all effects mentioned below.

1.3.1 Travel Time

If the distance between the Sun and Sirius is D, then light will cover it in time $T_l = D/c$. Using similar formula we get the travel time for the first spaceship

$$T_1 = D/v_1 = 3D/c = 3T_l,$$

and for the second spaceship

$$T_2 = D/v_2 = 2D/c = 2T_l.$$

Since the clocks of both spaceships were set to zero at the launch, the clock of the first spaceship will show $T_1 = 27$ years, and the clock of the second spaceship will show $T_2 = 18$ years.

Table 1.2 compares the travel times predicted by the Galileo-Newtonian mechanics to the results of the special theory of relativity. The clocks of the spaceships will show less time than intuitively expected. The last entry in the table (with the asterisk) corresponds to the case of a spaceship moving with the speed $v_* = 0.9c$. Notice that the travel time for this ship, according to its clock, is *less than 9 years!*

1.3.2 Relative Velocity

Relative to the Sun or Sirius, the second spaceship is moving with the speed $v_2 = 0.5c$. It is moving faster than the first spaceship, so the distance between the spaceships increases. But how fast? If the first ship measures the speed of the second ship, the result, based on Galileo-Newtonian physics, is simply

$$v_{21} = v_2 - v_1 = c/6.$$

Table 1.2 Time of travel for spaceships traveling from the Sun to Sirius, according to Galileo-Newtonian physics (GN) and the special theory of relativity (SR)

Time	GN	SR
1	27	25.5
2	18	15.6
*	10	4.4

Special relativity gives a different answer:

$$v_{21} = c/5.$$

As an extreme example, imagine a cosmic particle moving from the Sun towards Sirius with the speed $v_p = 0.999c$. It will pass by the second spaceship with the relative speed

$$v_{p2} = v_p - v_2 = 0.499c.$$

Special relativity says that the relative velocity is different and equals

$$v_{p2} = 0.998c.$$

1.3.3   Simultaneous Events

Imagine that at the moment of the launch two flares happen in two different locations: one on the Sun and the other in the Sirius system. The situation is illustrated in Fig. 1.5a. The flare from Sirius will reach the Sun 9 years after the

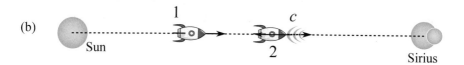

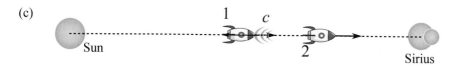

Fig. 1.5 Two flares happen in different locations at the moment of the launch. According to the observer in the solar system, the flares are simultaneous. What will the observers in the spaceships measure?

launch. Having taken into account the speed of light, the observer in the solar system concludes that both flares happened at the same time—they are *simultaneous*.

Neither of the spaceships will receive both flares simultaneously. The flare from the Sun will reach both ships right at the moment of the launch. The flare from Sirius will first reach the faster spaceship, moving with the speed $v_2 = c/2$, and then the slower one, moving with the speed $v_1 = c/3$; see Fig. 1.5b,c.

The time when the faster spaceship encounters the flare from Sirius is given by

$$t_2 = \frac{D}{c + c/2} = \frac{2}{3}T_l \approx 0.67T_l,$$

and similarly for the slower ship:

$$t_1 = \frac{D}{c + c/3} = \frac{3}{4}T_l = 0.75T_l.$$

According to the Galileo-Newtonian physics, once the ships take into account the propagation times t_1 and t_2, they will conclude that the flares on the Sun and Sirius happened at the same time—at the moment of the launch.

Special relativity tells us, firstly, that the clocks of the spaceships will show the following times when the flare from Sirius passes by them:

$$t_2 = \frac{1}{\sqrt{3}}T_l \approx 0.58T_l,$$

for the faster ship, and

$$t_1 = \frac{1}{\sqrt{2}}T_l \approx 0.71T_l.$$

for the slower ship.

Secondly, both spaceships will conclude that the flares *were not* emitted at the same time. If the first flare, the one at the Sun, happened at zero time, then the second flare must have happen *earlier* by

$$\tau_2 = T_l/2 \approx 4.5 \text{ years},$$

according to the faster spaceship, and by

$$\tau_1 = T_l/3 = 3 \text{ years},$$

according to the slower ship.

Remarkably, if the spaceships were moving with the same speeds but in the opposite direction (away from Sirius), they would have measured the flare from Sirius as happening *later* than the flare from the Sun.

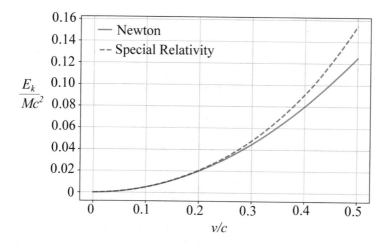

Fig. 1.6 Kinetic energy of a moving body according Newtonian mechanics (solid red line) and special relativity (dotted blue line). The energy is expressed in the units of Mc^2—the energy of a body at rest

1.3.4 Kinetic Energy

To accelerate a car with mass M up to the speed v, a certain amount of fuel must be spent.[7] The energy released by this fuel will be converted into the energy of the moving car—*kinetic energy*. In Newtonian mechanics, the formula for the kinetic energy is

$$E_k = \frac{Mv^2}{2}. \tag{1.1}$$

Special relativity gives a very different expression:

$$E_k = Mc^2 \left(\frac{1}{\sqrt{1 - v^2/c^2}} - 1 \right). \tag{1.2}$$

Despite their different look, the expressions (1.1) and (1.2) give close results for objects moving as fast as $c/3$, as shown in Fig. 1.6. To make the plot more readable, the velocity is given as a fraction of the speed of light. The kinetic energy for a body of mass M is given in the units of Mc^2—the energy of the body at rest.

---

[7] Here "fuel" means any kind of energy source used to move a car.

1.3.5 Momentum

Momentum is a quantity of great importance in physics. The law of momentum conservation is one of the most fundamental physical laws. For an object with mass M moving at the speed v, the momentum in Newtonian mechanics is given by

$$p = Mv. \tag{1.3}$$

In the special theory of relativity momentum has a different form:

$$P = \frac{Mv}{\sqrt{1 - v^2/c^2}}. \tag{1.4}$$

The conservation of momentum is extremely useful in studying particle physics. As every experimental particle physicist knows, the momentum given by (1.3) *is not conserved* in collisions of elementary particles, while the momentum given by (1.4) *is conserved* and therefore is the correct expression for momentum.

In Fig. 1.7 the expressions for momentum in Newtonian mechanics and in the special theory of relativity are compared, demonstrating that their values are close, as long as the speed of an object is small compared to the speed of light.

With the increasing speed of an object, the disagreement between Newtonian mechanics and special relativity becomes more prominent. Figure 1.8 illustrates the behavior of the kinetic energy and momentum at higher speeds ($v > 0.5c$).

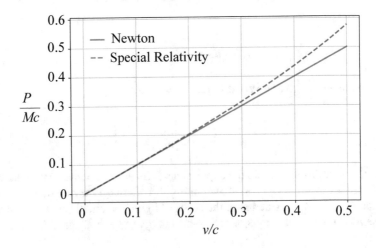

Fig. 1.7 Momentum of a moving body in Newtonian mechanics (solid line) and in the special theory of relativity (dotted line). For mathematical convenience only, momentum is given in the units of Mc— the Newtonian momentum of a fictitious body moving at the speed of light

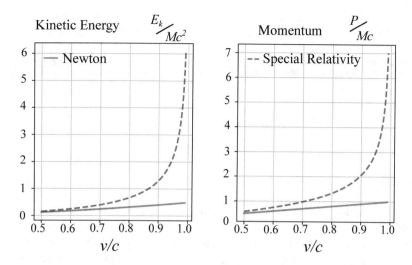

Fig. 1.8   Kinetic energy (left) and momentum (right) of a body moving at high speeds (0.5 < v < 1), according to Newtonian mechanics (solid red line) and special relativity (dashed blue line)

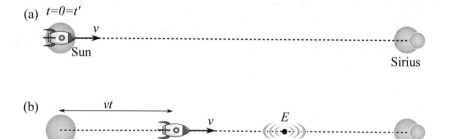

Fig. 1.9   An observer at rest in Solar system and an observer in a spaceship moving with speed v both measure position and time of the same event E

1.3.6   Transformation Equations

When two observers, moving relative to each other, observe the same event E and measure when and where it happens, they will get different results. This is true for Galileo-Newtonian physics and for the special theory of relativity.

Consider the situation in Fig. 1.9, where a spaceship is moving with the speed v relative to the Sun. Somewhere between the Sun and Sirius a collision of two asteroids happens, denoted as the event E.

The observer in the Solar system will find that the event E happen at time t and at location x, as measured from the Sun. The spaceship, flying by the Sun with the speed v, will measure the time t' and the location x' for the same event. According to Galileo-Newtonian physics, the connection between the measurements is simple:

▷ **Galilean transformation**

$$t' = t,$$

$$x' = x - vt.$$

The equations mean that, as long as both observers synchronize their clocks, they will assign the same time for the same event. The location of the event will only differ by an offset, equal to the distance traveled by the spaceship by the time when the event happens. The equations, mathematically expressing these assumptions, are called *Galilean transformation*.

Galilean transformation are inaccurate not only for large relative speed v, but also for small speed and large (e.g., astronomical) distances. In 1887 Woldemar Voigt discovered a different set of equations, which would be more accurate:

▷ **Voigt transformation**

$$t' = t - vx/c^2,$$

$$x' = x - vt.$$

Voigt transformation become inaccurate when the relative speed of observers is comparable to the speed of light. In this case the special theory of relativity should be used; it gives the exact relations, called *Lorentz transformation*:

▷ **Lorentz transformation**

$$t' = \frac{t - vx/c^2}{\sqrt{1 - v^2/c^2}},$$

$$x' = \frac{x - vt}{\sqrt{1 - v^2/c^2}}.$$

For small relative speed v and small distances x to events, the predictions of special relativity agree with Galileo-Newtonian physics. Lorentz transformation, the expression for momentum, kinetic energy, predictions about times and other measurements all reduce to the usual Galileo-Newtonian form.

Special relativity does not overthrow older physics, but extends it, bringing in agreement with all physical phenomena.

1.4 Who Needs Relativity?

In a daily life of a regular person the effects of special relativity are negligible. The physics of Galileo and Newton is sufficient to tackle many problems of mechanics. Who needs special relativity then?

Firstly, people dealing with objects moving with very high velocities: particle physicists and astrophysicists. There are many particle accelerators in the world. The largest particle collider today is Large Hadron Collider; it can accelerate protons up to 0.999999990 of the speed of light. Studying particles smashing into each other at high speeds allows high-energy physicists probe nature at the most fundamental level.

Fast moving particles are also created in space. The Sun, for example, produces energetic particles moving close to the speed of light. The universe appears to be full of processes that release huge amounts of energy: Quasars, gamma-ray bursts, colliding neutron stars, black holes devouring nearby stars. It is impossible to adequately describe these phenomena without special relativity.

Secondly, special relativity is important for people dealing with very high accuracy measurements of times and positions. The reliable functioning of Global Positioning System is impossible without taking special relativity into account. Another example is Scanning Electron Microscope (SEM)—an important scientific instrument, used to study objects at the scales down to nanometer (10^{-9} or 1/1000000000 of a meter). Scientists and engineers designing this instrument have to use special relativity to properly describe the motion of electrons used for imaging.

Finally, every physicist must learn special relativity, since it is the theory of fundamental physical notions—the measurements of time and lengths. Special relativity is especially important for those interested in general relativity. General relativity is currently the best theory of gravity with applications to astrophysics and cosmology.

1.5   Why Is Relativity Hard?

In his book *"The Meaning of it All"*,[8] the famous American physicist Richard
Feynman wrote:

*"Trying to understand the way nature works involves a most terrible test of
human reasoning ability. It involves subtle trickery, beautiful tightropes of logic on
which one has to walk in order not to make a mistake in predicting what will happen.
The quantum mechanical and the relativity ideas are examples of this."*

Scientists try to understand how nature works, with the goal of making useful
predictions and applications. In their experimental and theoretical work they rely on
"human reasoning ability". But how good is it? What are its limits?

We are looking at the universe using a tiny portion of the electromagnetic
spectrum. The smallest wavelength observed so far is about 10^{-12} nanometers
(10^{-21} of a meter), corresponding to a very high energy photon detected in 2019
coming from Crab Nebula. The radio waves have wavelengths on the order of 1000
meters (10^3 meters). Human eye is sensitive to electromagnetic radiation with the
wavelengths from 400 nm to 700 nm, roughly from 10^{-7} to 10^{-6} meters. The visible
range is just a tiny portion of the measured electromagnetic spectrum.

Looking at the world with a naked eye we can see objects as small as 0.1 mm
(10^{-4} or 1/10000 of a meter). For comparison, the size of a proton is about
1 femtometer (10^{-15} or 1/1000000000000000 of a meter). It is the gap in size
of 11 orders of magnitude—the difference between an individual person and the
population of 10 planets Earth. The proton, by the way, is not the smallest particle
because it has internal structure.

On the opposite end of the size scale is the observable universe, with its estimated
size of 4.4×10^{26} meters. Compare this to the Earth's highest mountain which is
roughly 10^4 meters high. Again, there is a large gap of 22 orders of magnitude
between cosmological distances and the distances available to our direct experience.

What about time? The blink of a human eye, which lasts about 0.1 of a second, is
often taken as the quickest process we register. The longest process we are familiar
with lasts about 100 years—the lifetime of a healthy human in the twenty-first
century. The span of 10 orders of magnitude (from 10^{-1} second for an eye-blink
to 10^9 seconds for 100 years) is impressive.

In ultra-fast optics, physicists are able to study processes with durations as small
as 100 femtoseconds (10^{-13} or 1/10000000000000 of a second). This is 12 orders
of magnitude faster than we can imagine. On the opposite side of the time scale is
the age of the universe—14 billion years (4.4×10^{19} seconds). Once more, a huge
gap of 10 orders of magnitude between our direct experience and the longest natural
phenomenon.

---

[8] R. Feynman, *The Meaning of It All*, Addison-Wesley, 1998.

Similar comparison can be made for our familiarity with various speeds. Even if we hop on the fastest supersonic jet, we will be moving billion times slower than light or many cosmic particles.

All these comparisons demonstrate how limited our experience is, compared to what the world has to offer. Our common sense and intuition are based on such a limited foundation that it is surprising that we managed to understand so much about the universe. Talking about the world around us in " *Physics and Reality*",[9] Albert Einstein, said: *"The fact that it is comprehensible is a miracle."*

The special theory of relativity tells us how world works at high speeds, high energies, long times, and large distances. We have no experience in these ranges, our intuition fails and often gets in a way. This is one of the reasons why relativity is hard. To make it easier, it is better not to cling to usual notions of time and lengths, momentum and energy.

If we can not use common sense and intuition, what can we rely on? The best tool science has found so far is abstract thinking and mathematics. Fortunately, the mathematics of special relativity is not complicated. Some high-school algebra, trigonometry, rudiments of calculus, and basic analytic geometry are sufficient.

Chapter Highlights

- The special theory of relativity emerged as the result of work of many scientists. Lorentz, Poincare, and Einstein are among the most prominent contributors into the theory. History of the special theory of relativity is rich and nuanced.
- The important concepts of the theory of relativity include observers and the laws of physics. There are many observers that discover the same laws of physics. Such observers are equivalent, despite being different as regards their relative position, orientation, and the state of motion.
- The special theory of relativity does not overthrow the previous theories of physics, but corrects and extends them into new domains of energies, velocities, and distances. The limited human intuition often finds it hard to accept the consequences of the special theory of relativity, despite their continuous success and experimental confirmations.
- The special theory of relativity is an important part of physical picture of reality. It is an essential element in the education of every physicist.

References

1. Feynman, R. (1955). The value of science. In *Address given at the 1955 meeting of the National Academy of Sciences*.
2. Miller, A. (1998). *Albert Einsteiňs special theory of relativity*. Springer.

---

[9] A. Einstein, *Physics and Reality*, Daedalus, Vol. 132, No. 4, On Science (Fall, 2003), pp. 22–25.

3. Galilei, G. (1953). *Dialogue concerning the two chief world systems* (pp. 186–187), translated by Stillman Drake, University of California Press.
4. Poincare, H. (1913). *The principles of mathematical physics* (pp. 297–320). Translation in The Foundations of Science (The Value of Science). Science Press, Chap. 7–9.
5. Feynman, R. (1998). *The meaning of it all*. Addison-Wesley.
6. Einstein, A. (2003). Physics and reality. *Daedalus, 132*(4), 22–25, On Science (Fall).

Chapter 2
Geometry

> It is meant that by natural selection our mind has adapted itself
> to the conditions of the external world, that it has adopted the
> geometry most advantageous to the species: or in other words
> the most convenient. This is entirely in conformity with our
> conclusions; geometry is not true, it is advantageous.
>
> H. Poincare, *Science and Hypothesis, Ch V: Experience and
> Geometry*

Abstract Geometry plays an important part in the mathematical description of the special theory of relativity. The concepts of points, figures, geometric laws, and coordinate systems have illuminating counterparts in physics. Additionally, the important distinction between *relative* and *invariant* quantities, essential in the special theory of relativity, naturally appears even in planar Euclidean geometry. This chapter discusses main geometrical ideas and tools that help understand special relativity.

2.1 What Is Geometry?

According to the modern view of geometry, put forward in 1872 by a German mathematician Felix Klein, the study of geometry requires three main components: *points*, *figures*, and *operations*.

1. *Points* represent the most primitive elements of geometry. The totality of all points, considered as one whole, represents *space* in the mathematical sense.
2. *Geometrical figures* are "built" from points. They are collections (sets) of points that occupy parts of space.
3. A geometric *operation* takes a figure and produces another one, possibly shifted, rotated, scaled, or otherwise *transformed*. Operations are sometimes called *transformations*.

© The Author(s), under exclusive license to Springer Nature Switzerland AG 2022
Y. Deshko, *Special Relativity*, Undergraduate Lecture Notes in Physics,
https://doi.org/10.1007/978-3-030-91142-3_2

In planar Euclidean geometry we deal with regular points of a plane; geometric figures, such as lines, circles, rectangles, etc.; and with operations, such as rotations, reflections, and shifts (translations).

Like many ideas in modern mathematics, Felix Klein's idea of geometry is abstract and general. The motivation for such an abstract view was to bring under the same roof a number of mathematical theories that showed similarities with Euclidean geometry, but were not Euclidean. Non-Euclidean geometries were actively studied in the nineteenth century, and that created favorable conditions for the discovery of the theory of general relativity as the geometrical theory of gravity.

Once all three components are defined, we can study different *properties* of geometrical figures. In planar geometry the results are expressed as theorems, such as the law of sines, the parallelogram law, Thales's theorem, and many others. The following exercise gives another example of a geometric law.

? Problem 2.1

Prove that angles α and β with perpendicular sides are equal. See Fig. 2.1.

The laws of planar geometry involve lengths, angles, or areas of figures. There is a profound reason for that. According to Klein's view, a geometrical property of a figure *must not change* when a geometrical operation is applied to it. For example, the sum of angles in a triangle remains the same when it is translated, rotated, or flipped. Discovery of such *invariant* properties is the goal of geometry.

Fig. 2.1 Two angles with perpendicular sides are equal: $a \perp b$, $a' \perp b'$, and $\alpha = \beta$

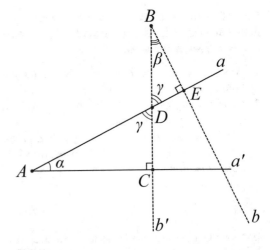

As an example of a non-invariant property, consider the statement about the lower left triangle $\triangle ACD$ in Fig. 2.1: *Right angle $\angle C$ is below the hypotenuse AD.* This statement is true for the triangle in Fig. 2.1, but it becomes false if we rotate the triangle by $180°$.

The meaning of geometry can be summarized as follows:

▷ **Meaning of Geometry**

Geometry studies invariant properties of geometrical figures. Properties must be invariant with respect to certain operations or transformations. Points make up geometrical figures and spaces. For different definitions of points, used in mathematics and physics, there are different meaning for figures, operations, invariant properties and, consequently, different geometries. Euclidean geometry is one of many possible geometries.

2.2   Coordinates

Geometrical properties of figures do not depend on whether we label their points; parts of figures are labeled simply for convenience. In the Exercise 2.1 (see Fig. 2.1) letters from Greek and Latin alphabet were used to label some points and angles.

Coordinates provide a systematic way to label *any* point with a number (or several numbers), uniquely identifying it among the multitude of other points. Introducing numerical labels through coordinates creates a bridge between algebra and geometry and leads to a powerful approach to geometric problems—*algebraic geometry*.

There are many ways to introduce coordinates. In planar geometry two often used systems are *Cartesian* and *polar*.

2.2.1   Cartesian Coordinates

Renes Descartes, a French scholar who lived in the seventeenth century, suggested the following simple way to construct a *coordinate system* in a plane:

1. Choose an arbitrary point O, as shown in Fig. 2.2. It will become the *origin* of the coordinate system.
2. Draw a straight line through the origin O. This is the first *coordinate axis*. The direction of this axis is arbitrary and should be chosen to suit specific problem.
3. Draw another straight line through the origin O, at a right angle to the first axis. This is the second *coordinate axis*.

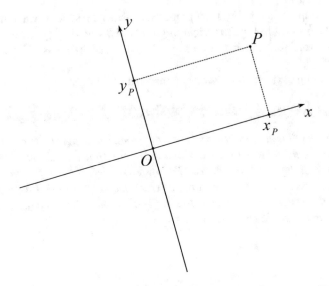

Fig. 2.2 Cartesian coordinate system needs a point of origin O, and two perpendicular axes, x and y. The position of the origin and the direction of the axes are arbitrary

The first axis is often drawn horizontally and called x axis; the second axis is then drawn vertically and called y-axis.

Once the origin and the axes are fixed, we can assign a label to any point P in the plane in two steps (see Fig. 2.2):

1. Starting at P, draw a straight line parallel to the y-axis until it intersects the x-axis at the point x_P. The distance from the origin O to x_P will be the first part of the label. Since the same distance from O may be measured in two different directions, we can call one direction positive, and express all distances measured in this direction using positive numbers. All distances measured in the opposite direction will become negative.
2. Starting at P, draw a straight line parallel to the x-axis until it intersects the y-axis at the point y_P. The signed distance from the origin O to y_P will be the second part of the label. The choice of the sign follows the same rule as for the x axis.

This way a pair of *Cartesian coordinates* can be assigned to any point in a plane. Symbolically, this fact is expressed as

$$P \rightarrow (x_P, y_P) \quad \text{or} \quad \bullet \rightarrow (x, y).$$

The choice of coordinate axes is arbitrary; it is often dictated by the problem at hand. The following exercise demonstrates how the apparent difficulty of a problem depends on the orientation of the axes relative to a figure.

Find the area of a figure specified by the following inequalities in Cartesian coordinates:

(A)

$$|x| \le 1$$
$$|y| \le 1. \tag{2.1}$$

(B)

$$|x + y| \le \sqrt{2}$$
$$|x - y| \le \sqrt{2}. \tag{2.2}$$

(C)

$$|\sqrt{3}y + x| \le 1$$
$$|y - \sqrt{3}x| \le 1. \tag{2.3}$$

2.2.2   Polar Coordinates

The way to introduce *polar* coordinates in a plane is illustrated in Fig. 2.3. It requires two steps:

1. Choose an arbitrary point O, called a *pole*, which serves as the origin of the polar coordinate system.
2. Draw a straight line, starting at the origin O and going in an *arbitrary* direction. This line, called a *polar axis*, is used as a reference for measuring angles.

Once the pole and the polar axis are chosen, we can assign a two-number label to any point P in the plane. The first number is the distance r_P from the pole O to the point P; the second is the angle ϕ_P between the line OP and the polar axis. Since the same angle may correspond to two different points located on the opposite sides of the polar axis, we adopt a *convention for measuring angles*:

▷ **Convention on Angles**

If an angle is measured counter-clockwise – it is positive. A clockwise measured angle is negative.

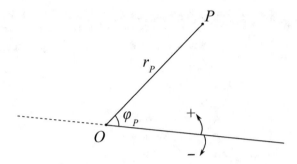

Fig. 2.3 Polar coordinate system requires a pole O and an axis. Any point P is labeled with two numbers: the distance r_P from the pole and the angle ϕ_P, measured from the axis. Angles measured counter-clockwise from the polar axis are considered positive, otherwise—negative. In this example, $\phi_P > 0$

Using polar coordinates, a pair of numbers can be assigned to any point in a plane. Symbolically this is expressed as

$$P \to (r_P, \phi_P) \quad \text{or} \quad \bullet \to (r, \phi).$$

Isaac Newton was one of the first scientists to use polar coordinates to analyze curves, such as parabolas and spirals, in a plane. See C. B. Boyer "Newton as an Originator of Polar Coordinates" in *The American Mathematical Monthly*, Vol. 56, No. 2.

2.2.3   *Arctan Coordinates**

Both Cartesian and polar coordinates cover the whole plane. The points farther from the origin get larger coordinates, and the points at infinity get infinite coordinates x, y and r. It is possible, however, to cover an infinite plane with finite coordinates for all points. To see this, we can consider a simple case of an infinite straight line, as shown in Fig. 2.4.

First, choose an origin O on the line and draw a circle with radius R touching the line at the origin. Then pick a point P on the line, and connect it with the center of the circle C. The line connecting the point P and the center of the circle C will

---

*Chapters or sections marked with an asterisk sign contain advanced material. They can be skipped if one looks for a shorter introduction into the special theory of relativity.

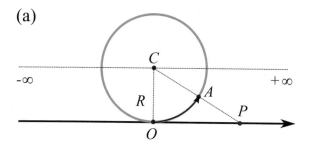

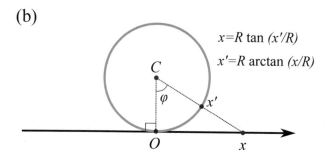

Fig. 2.4 A method of covering an infinite line with coordinates from a finite range $[-\pi/2, \pi/2]$. It relies on "mapping" points on the line into points on the circle. All points on the line, including the infinitely far removed from the origin, receive finite coordinates

intersect the circle at the point A. The length of the arc OA is the arctan coordinate of the point P, which we will denote x'.

For points approaching infinity on the right (left) side of the origin O the coordinates will approach the finite value $\pi R/2$ (or $-\pi R/2$ for the left side). Using the triangle OCP in Fig. 2.4a, we find the connection between the coordinate x of a point P measured along the straight line, and the coordinate x' measured along the circle. The length of the arc is

$$x' = R\phi .$$

From the definition of the trigonometric function $\tan\phi$, we find

$$\tan\phi = \frac{x}{R} \quad \rightarrow \quad \phi = \arctan\frac{x}{R},$$

leading to

$$x' = R\arctan\frac{x}{R}.$$

For a circle with radius $R = 1$ this reduces to

$$x' = \arctan x.$$

In a plane, a similar procedure can be used for both axis x and axis y. The result is a pair of finite coordinates for any point in a plane:

$$x' = \arctan x,$$
$$y' = \arctan y. \tag{2.4}$$

Using such coordinates allows "squeezing" (or *mapping*, in mathematical lingo) the entire infinite plane into a finite patch of the plane. Arctan coordinates are used in Penrose diagrams, as explained in the Appendix B.3.

2.2.4 Cantor Coordinates*

In 1878, a German mathematician Georg Cantor proved the following curious fact: *For every point on a line there is a corresponding and unique point in a plane, and vice versa.* Such a one-to-one correspondence means that any point in a plane can be labeled with *just a single number*, as opposed to a pair of numbers, used in Cartesian or polar coordinates.

To understand how this works, consider a point P in a plane with Cartesian coordinates x_P and y_P, as shown in Fig. 2.5. For any point P inside the square $0 \leq x, y < 1$, the coordinates can be written using decimal point representation:

$$x_P = 0.x_1 x_2 x_3 \dots ,$$
$$y_P = 0.y_1 y_2 y_3 \dots . \tag{2.5}$$

A simple way to replace these two numbers with just one is to "zip" them, writing

$$z_P = 0.x_1 y_1 x_2 y_2 x_3 y_3 \dots \tag{2.6}$$

This way we get a *unique* number z_P, and can extract the original numbers x_P and y_P also in a unique way, by an obvious procedure of "unzipping." Cantor's method shows that just one axis is sufficient to label points in a plane. If all we need is to label points, then Cantor's method gives a valid coordinate system.

This approach can be generalized to points in spaces of three and more dimensions. Combined with the arctan coordinates, Cantor coordinates demonstrate that a line segment with finite length is sufficient to label *any point in a space of any finite dimensions.*

Fig. 2.5 Labeling of points
in a plane using Georg
Cantor's method. If, for
example, the point P has
Cartesian coordinates
$x = 0.64937458$, and
$y = 0.94027305$, then we can
replace a pair (x, y) with a
single number
$z = 0.6944903277435085$.
The original coordinates x
and y are unambiguously and
uniquely extracted from z

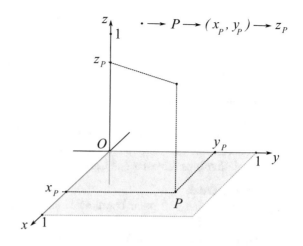

2.2.5   Remarks on Coordinates

Cartesian and polar coordinate systems are easy to construct and their coordinates
have clear meaning. In contrast, Cantor coordinates are defined using coordinate
transformation; it's not obvious what kind of measuring device can produce these
coordinates. Many more useful coordinate systems are possible, but not always
coordinates have a simple intuitive meaning.

There is a lot of freedom in the choice of coordinates. It is important to remember
that *the laws of geometry do not depend on the choice of the coordinate system.*
The freedom, however, is not absolute, since we expect the coordinates to behave
"reasonably" in the following sense:

1. Every point gets a unique set of coordinates.
2. For two very close points their coordinates are also very close.

Cartesian coordinates provide an example of such well-behaved coordinates. In
polar coordinates the origin is singled out among all other points. Any pair of
coordinates $(0, \phi)$ corresponds to the origin of the polar coordinate system, the
requirement of uniqueness is thus not satisfied.

As an extreme case, consider a coordinate system where points are assigned
labels randomly. It may happen that two distinct points will get the same coordinates, or two very close points will have a large difference of their coordinates.
These are examples of coordinate irregularities that do not reflect properties of
Euclidean plane; coordinate systems with such irregularities should be avoided.

> Cantor coordinates are not well-behaved and are not used in physics. They
> remind us that we must be careful when exercising the freedom of choice of
> the coordinate system.

2.2.6   *Transformations of Coordinates*

It may be useful to consider the same problem in different coordinates. When two
coordinate systems are used at the same time, any point will have two sets of
coordinates. Table 2.1 lists several points, $P, Q, R, \ldots$, with the coordinates (x, y)
in some coordinate system CS, and the coordinates (x', y') in another coordinate
system CS'.

The relationship must exist between the coordinates (x, y) and (x', y'). The
arctan coordinates (2.4) provide an example of such a relationship. We can find
similar relationships for polar and Cartesian coordinates.

Polar to Cartesian

Consider Cartesian and polar coordinates sharing a common origin and axis, as
shown in Fig. 2.6.

Looking at the right triangle $\triangle OPQ$ in Fig. 2.6 and using the definition of
trigonometric functions $\sin \varphi$ and $\cos \varphi$, we get

$$x_P = r_P \cos \phi_P,$$

$$y_P = r_P \sin \phi_P.$$

These equations express Cartesian coordinates (x_P, y_P) of any point P, given its
polar coordinates (r_P, ϕ_P). For an arbitrary unnamed point the equations have the
form without subscripts:

Table 2.1 The same point •
gets a unique set of
coordinates (x, y) in a
coordinate system CS. In
another coordinate system
CS', it gets another unique
set of coordinates (x', y')

CS	Point	CS'
(x_P, y_P)	$\leftarrow P \rightarrow$	(x'_P, y'_P)
(x_Q, y_Q)	$\leftarrow Q \rightarrow$	(x'_Q, y'_Q)
(x_R, y_R)	$\leftarrow R \rightarrow$	(x'_R, y'_R)
...	...	...
(x, y)	$\leftarrow \bullet \rightarrow$	(x', y')

Fig. 2.6 Transformation of
coordinates from polar into
Cartesian can be found by
looking at the right triangle
$\triangle OPQ$ and using
trigonometric functions:
$x_P = r_P \cos \phi_P,$
$y_P = r_P \sin \phi_P$

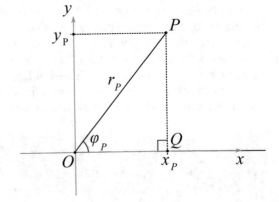

$$x = r \cos \phi, \tag{2.7}$$

$$y = r \sin \phi. \tag{2.8}$$

Find the equations for polar coordinates in terms of Cartesian.

Cartesian to Cartesian

The same geometric problem can be analyzed in two different Cartesian coordinate systems. Since two sets of axes and origins are used, we need to denote them differently. The first Cartesian coordinate system, named S, is denoted as $S = \{O, x, y\}$, and the second as $S' = \{O', x', y'\}$. This notation emphasizes the fact that Cartesian coordinate system requires an origin and two axes.

There are three basic ways in which S and S' may differ; they are illustrated in Fig. 2.7:

1. Axes x, x' and y, y' have the same orientations, but the origins O and O' are shifted.
2. Origins O and O' coincide, but the x' axis is rotated relative to the x axis by some angle θ. The y' axis must also be rotated by the same angle θ relative to the y axis, since x' and y' are rigidly fixed to each other, making the right angle.
3. Origins O and O' coincide, but the direction of one of the axes (either x' or y', *but not both*) is flipped. Flipping the directions of both axes is a special case of rotation by $\pm\pi$.

Any combination of the basic operations listed above is also possible.

The first and the third cases are easy to analyze; below we will study only the rotation of the axes. Figure 2.8 illustrates the steps necessary to find the relationship between the Cartesian coordinates (x, y) and (x', y') of the same point in the coordinate systems S and S', respectively.

To avoid clutter, we first consider what happens to the coordinate x'_P of an arbitrary point P when the axis x' is rotated by an angle θ counter-clockwise, as shown in Fig. 2.8a. The right triangles $\triangle OPx_P$ and $\triangle OPx'_P$ have common hypotenuse, and the angles α and β between the hypotenuse and the axes x and x' are related:

$$\alpha = \beta + \theta,$$

where θ is the rotation angle of the coordinate axes.

The distance to the point x'_P is

$$x'_P = r \cos \beta = r \cos(\alpha - \theta).$$

Fig. 2.7 Three basic ways in
which two Cartesian
coordinate systems may
differ: (**a**) Shift of the origin;
(**b**) rotation around the origin;
(**c**) inversion of one of the
axes

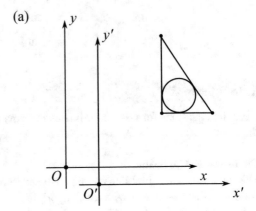

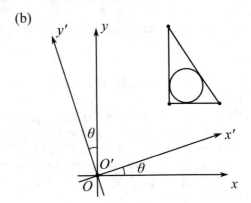

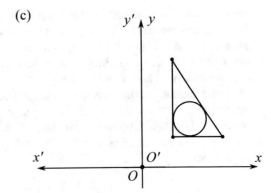

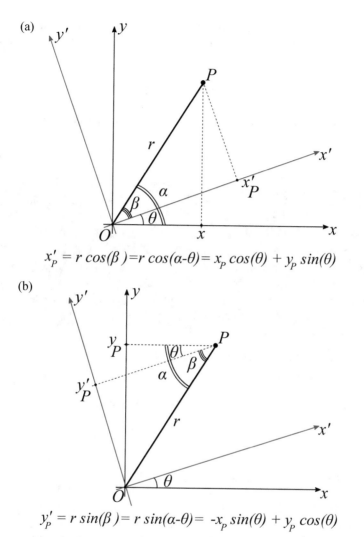

(a)

$$x'_P = r\,cos(\beta\,)=r\,cos(\alpha\text{-}\theta)= x_P\,cos(\theta) + y_P\,sin(\theta)$$

(b)

$$y'_P = r\,sin(\beta\,)= r\,sin(\alpha\text{-}\theta)= \text{-}x_P\,sin(\theta) + y_P\,cos(\theta)$$

Fig. 2.8 Transformation of the coordinates x and y of a point P from one Cartesian coordinate system S into another, S'; the latter is rotated relative to S by an angle θ

Using the trigonometric identity

$$\cos(a + b) = \cos a \cos b - \sin a \sin b,$$

and the fact that $\sin(-a) = -\sin a$, the coordinate x'_P is written

$$x'_P = r \cos \alpha \cos \theta + r \sin \alpha \sin \theta.$$

The factors $r \cos \alpha$ and $r \sin \alpha$ can be recognized as the coordinates x_P and y_P, respectively (see the relationship between polar and Cartesian coordinates, discussed above). We then find

$$x'_P = x_P \cos \theta + y_P \sin \theta. \tag{2.9}$$

Transformation of the y'_P coordinate can be found in a similar way:

$$y'_P = -x_P \sin \theta + y_P \cos \theta. \tag{2.10}$$

The position of a point P is arbitrary. In the derivation given above, it was placed in the top right part of the (x, y)-plane. For any other position the derivation will be similar and the result for the coordinate transformation will be the same.

In fact, if the point P is placed in any other *quadrant* of the (x, y) plane, we could do one, two, or three, if required, rotations of the axes by $\pi/2$ first to bring the point P into the top right part. The combination of such rotations with the rotation by θ will result in the desired transformation equations.

Coordinate transformations we derived so far are summarized in Table 2.2.

? Problem 2.4

Derive the formulas for the coordinate transformations from $S' = \{O', x', y'\}$ to $S = \{O, x, y\}$.

? Problem 2.5

Using the transformation relationship between (x', y') and (x, y), show that when switching between the Cartesian coordinates systems $S = \{O, x, y\}$ and $S' = \{O', x', y'\}$, the distance between two points P and Q remains the same. The distance between points P and Q is given by

$$D = \sqrt{(x_Q - x_P)^2 + (y_Q - y_P)^2}$$

Table 2.2 Transformation of coordinates between polar and Cartesian, between two different Cartesian coordinate systems; and between Cartesian and arctan coordinates

Polar to Cartesian	Cartesian S to Cartesian S'	Cartesian to arctan
$x = r \cos \phi$	$x' = x \cos \theta + y \sin \theta$	$x' = \arctan x$
$y = r \sin \phi$	$y' = -x \sin \theta + y \cos \theta$	$y' = \arctan y$

in S, or

$$D = \sqrt{(x'_Q - x'_P)^2 + (y'_Q - y'_P)^2}$$

in S'.

2.3   Spaces and Dimensions*

2.3.1   Space

Space is where stuff is; there can be lots of space or not enough space; space can be occupied or empty. Space is a simple idea as long as we do not think about it too much. But in the special theory of relativity we need to study the notion of space more carefully. So what is space in mathematics and physics?

Space in mathematical sense is a rich and powerful idea. Starting with familiar three dimensional space with Euclidean geometry, mathematics and physics significantly extend this notion to include many more interesting and important concepts. The price to pay is that the idea of space becomes more *abstract*.

▷ **Spaces in Mathematics and Physics**

Here is a short list of various spaces used in mathematics and physics:

Vector space; *topological* space; *affine* space; *metric* space; *projective* space; *Banach* space; *Hilbert* space; *Hausdorff* space; *Fock* space; *Minkowski* space; *configuration* space.

This list illustrates how diverse the notion of space is in modern science. Later we will study Minkowski space in details.

Let us consider several examples of spaces used in mathematics and physics, to see their similarities and differences.

Euclidean Space
Whenever you find yourself in a room, you may build a mental picture of a three dimensional Euclidean space. A position of any point inside the room can be specified using three Cartesian coordinates, as shown in Fig. 2.9. The origin can be placed in any corner and the directions of the axes are then easily defined by the lines where the walls meet the floor.

The distance from the origin O to any point P can be measured directly, or using the shortest distances from the point P to both walls and the floor. For this space,

Fig. 2.9 Three dimensional
Euclidean space is a
mathematical representation
of usual physical space we
observe around us

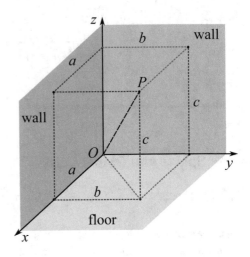

Pythagoras' theorem holds:[†]

$$OP^2 = a^2 + b^2 + c^2. \qquad (2.11)$$

Every point inside the room is uniquely identified by a set of three numbers
(a, b, c)—its Cartesian coordinates. All points, taken together as a whole, make
Euclidean space—the space where the geometry of Euclid is valid. Let's summarize
this important example:

▷ Euclidean Space

*In Euclidean space there are points, figures, such as lines, triangles, circles, cubes,
etc. The relations between points or figures involve distances and angles. The laws
of Euclidean geometry are true for figures in Euclidean space.*

Polynomial Space

Consider a quadratic polynomial in a variable v:

$$3v^2 - 4v + 7.$$

How many polynomials of this kind can we write?

The coefficients in quadratic polynomial can be any numbers, so we can write
the general expression as follows:

[†] It follows from two successive application of Pythagoras' theorem: first to find the distance
squared of the diagonal line on the floor, and then the distance squared of the segment OP.

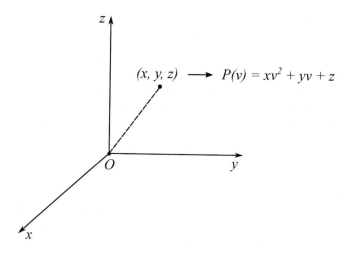

Fig. 2.10 Any quadratic polynomial $P(v)$ can be viewed as a *point* in the space of all possible quadratic polynomials—*polynomial space*

$$x v^2 + y v + z. \tag{2.12}$$

Three numbers (x, y, z) uniquely specify a quadratic polynomial. They also specify a unique point in a Euclidean space. We thus have a one-to-one correspondence between points in three dimensional Euclidean space and quadratic polynomials

$$P(v) = x v^2 + y v + z.$$

The correspondence is illustrated in Fig. 2.10.

It is important to realize that when talking about a quadratic polynomial $P = x v^2 + y v + z$, we are not concerned with its graphical representation as a curve in a plane (parabola). It is the expression for P that we call a point, and a variety of such expressions we consider as a set of points.

It is possible to view a single polynomial $P(v)$ as a *point*, a basic and simplest element, a mathematical "atom". It is not literally a point that we are used to think about and point a finger to; however, it can correspond to a point in Euclidean space. All possible quadratic polynomials make a *polynomial space*.[2]

? Problem 2.6 **

Consider a geometric figure in quadratic polynomial space, defined by the following condition: $P(v) = 0$ has only one solution. First, find all polynomials that satisfy this requirement. Second, visualize the resulting figure in polynomial space (x, y, z).

State Space

In physics, the behavior of various gases is studied by measuring their pressure P, volume V, temperature T, and the number of particles N (see Fig. 2.11). For a known quantity of gas, the temperature can be determined using the ideal gas equation:

$$PV = NkT,$$

where k is a fundamental physical constant, called Boltzmann constant; this constant translates between temperature in degrees and energy in Joules.[3]

   The minimal number of *parameters* needed to describe the properties of an ideal gas is three: P, V, and N (the temperature can be found from the ideal gas equation). The triple (P, V, N) defines the *state* of the gas.

Fig. 2.11 The state of an ideal gas is determined by measuring its pressure P, volume V, and the number of gas particles N. From these three numbers (P, V, N) the temperature can be calculated using the ideal gas equation: $PV = NkT$, where k is the fundamental physical constant (a number), called *Boltzmann constant*

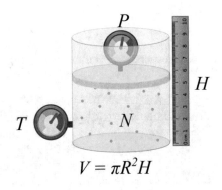

$$V = \pi R^2 H$$

---

[2] A more precise term would be *quadratic polynomial space*.

**Problems marked with an asterisk sign are more challenging than the rest. Double asterisk sign marks an especially challenging problem

[3] $k = 1.38064852 \times 10^{-23}$ J/K.

Fig. 2.12 A figure in the gas state space corresponds to some thermodynamic process

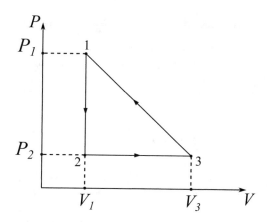

Table 2.3 Some examples of spaces used in physics and mathematics

Space	Point
Euclidean space	Place or location
Polynomial space	Polynomial equation $P(v)$
Gas state space	Physical state of an ideal gas

Three numbers (P, V, N) are analogous to three Cartesian coordinates (x, y, z) of some point in Euclidean space. In Euclidean space Cartesian coordinates range from negative infinity to positive infinity; however the coordinates of the state of a gas are all positive. Also, each coordinate has different physical meaning: P is the pressure, measured in Pascals; V is the volume, measured in cubic meters; T is the temperature, measured in degrees. Nevertheless, we can view the state of a gas as a *point* in *state space*—collection of all physically possible states.

? Problem 2.7

Consider a gas with a known number of particles N. It is initially in a state 1 with the pressure P_1 and the volume V_1, as illustrated in Fig. 2.12. The gas then goes through a *thermodynamic process* 1-2-3-1.

Describe what happens to the gas during each step 1–2, 2–3, and 3–1. What are the temperatures of the gas in the states 1, 2, and 3?

A short list of spaces, considered above, is summarized in Table 2.3.

By now, the idea of a general space should be clear:

Space is a collection of points. A point is not always a location that we can point to, but simply the most primitive element or fundamental building block.

2.3.2   Dimensions

To specify a point in a plane using polar or Cartesian coordinates, two numbers are required. One coordinate is not enough, three coordinates for the plane will be too many, and it looks like precisely $N = 2$ numbers are needed. For the space around us we need $N = 3$ numbers. This leads to the idea of a *dimensionality* of a space. At the first glance, the number of coordinates used to locate a point in space defines its dimensionality. However, the example of Cantor's coordinates shows that we must be careful. The minimal number of coordinates required to label any point in any space of finite dimensions is one. Does this mean that the dimension of any space is $N = 1$? Turns out, there are at least two more ways to determine the dimensionality of a space.

Dimensionality from Boundaries

The first approach is based on three simple rules:

1. Dimensionality of a point, or any countable number of points, is zero.
2. Dimensionality of an infinite space is the same as dimensionality of its finite bounded regions.
3. Dimensionality of a bounded region of space equals the dimensionality of its boundary plus 1.

The application of the method of boundaries is illustrated in Fig. 2.13 for the case of a plane. To determine the dimensionality of the plane, we first select a part of the plane (region). The boundary of this region is a curve; it is a closed curve with no boundary. To find the dimensionality of the curve, we again select a part of the curve (a bounded region). A region of the curve is bounded by two points. Provided that the dimensionality of two points is zero, we can work backwards to deduce that the dimensionality of the curve is one, and the dimensionality of the plane is two. Note that no coordinates were used in this method.

Dimensionality from Directions

The second approach to determine the dimensionality of a given space can be used if the space possesses the notion of a *direction*. Figure 2.14 shows how this is done for a plane.

Starting from an arbitrarily selected origin O, we can draw an arrow $\mathbf{a}_1$ in *any* direction. The second arrow $\mathbf{a}_2$ can point anywhere, except in the direction parallel to $\mathbf{a}_1$, i.e. neither in the same nor opposite direction. This exhausts all possible *independent* arrows.

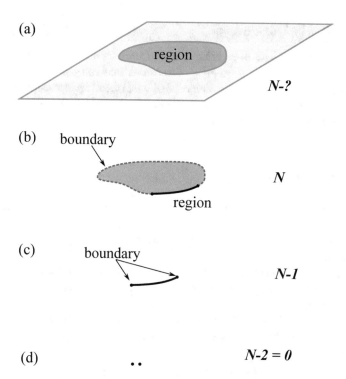

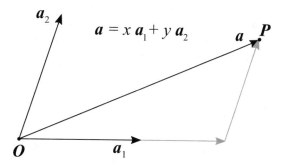

Fig. 2.13 To find the dimensionality of a space (plane in this example) we can use recursive definition based on the dimensionality of the boundary of some part of this space. Dimensionality of a point is zero *by definition*

Fig. 2.14 To find the dimensionality of a space (plane in this example) we can use the number of *independent directions*

An arbitrary arrow **a** will connect the origin and some point P (see Fig. 2.14). The direction along OP is not independent from the directions specified by the arrows $\mathbf{a}_1$ and $\mathbf{a}_2$ in the following sense: The point P can be reached in two consecutive steps. First, move along $\mathbf{a}_1$ some distance x, then along $\mathbf{a}_2$ some distance y. Clearly, using a single direction $\mathbf{a}_1$ is not enough to reach *any* point in the plane, while using two is just enough; using three is redundant. The plane is two-dimensional according to this criterion.

> ▷ **Dimensionality of Space**

Nearly all spaces used in physics and mathematics have well defined dimensionality. To determine the number of dimensions one can use either the number of independent directions or the method of boundaries. Most coordinate systems are "well-behaved" and the number of coordinates gives the same number of dimensions as the other two methods.

Chapter Highlights

- The ideas of space and points are generalized in modern mathematics and physics.
- The essence of geometry is the study of various *relationships between points* and figures in a given space.
- The relationships between points and figures which are *invariant* (i.e. remain the same) when we "change our point of view" on the space (apply a transformation) are of central importance.
- The use of coordinate systems to label points in space is not mandatory, but helpful; it brings the mathematical tools of algebra and calculus into the domain of geometry.
- There exists a great freedom in the choice of coordinate systems in a given space.
- Most coordinate systems are well-behaved and the number of coordinates, necessary to locate a point, determines the dimensionality of space. Additionally, the methods of independent directions and the recursive method of boundaries can be used to find the number of dimensions.
- Different spaces are characterized by different kinds of relationships between their points; hence different kinds of geometries. Euclidean geometry is one of many possible geometries; it is one of the simplest.

References

1. Poincare, H. (1905). Experience and geometry. In *Science and hypothesis*. Science Press, Ch V.
2. Boyer, C. B. (1949). Newton as an originator of polar coordinates. *The American Mathematical Monthly*, 56(2), 73–78.

Chapter 3
Spacetime

It is a widespread error that the special theory of relativity is supposed to have, to a certain extent, first discovered, or at any rate, newly introduced, the four-dimensionality of the physical continuum. This, of course, is not the case. Classical mechanics, too, is based on the four-dimensional continuum of space and time.

A. Einstein, *Autobiographical Notes: A Centennial Edition, Open Court, LaSalle and Chicago, 1979*

Abstract Building upon the idea of space as a collection of points, we study a special space used in relativity, called *spacetime*. The points of this space are physical *events*. Starting with the definition of an *event*, we will discuss how the relations between different events, such as time interval and distance, are measured. The measurements of times and distances lead to the notion of the *frame of reference*, which serves as a coordinate system in spacetime. We will discuss how to represent spacetime and events graphically using *spacetime* diagrams. This will further the geometrical approach to the theory of relativity.

3.1 Events

From planets going around the sun to elementary particles accelerating in colliders, physics studies, among other things, *how* and *why* objects move. One can measure the object's position, velocity, and acceleration as they change in time. In this approach physical *objects* are the basic elements and come into the picture first.

In the alternative approach, *events* that occur are viewed as more fundamental or primitive elements. After all, we perceive the world through events; groups of events related in some sense give rise to the idea of various objects.[1]

Here are several examples of events:

---

[1] This question is discussed in more details in the Appendix B.5.

© The Author(s), under exclusive license to Springer Nature Switzerland AG 2022
Y. Deshko, *Special Relativity*, Undergraduate Lecture Notes in Physics,
https://doi.org/10.1007/978-3-030-91142-3_3

- An atom emits or absorbs a light particle (photon).
- Two protons collide.
- An electron and positron annihilate.
- A radioactive atom emits a neutrino.
- A bullet hits a target.
- A flashlight is turned on.
- A spacecraft receives a signal.

It may be asked: How can an event (e.g., a collision of protons) happen without objects (protons) existing first? This is not a "chicken or the egg" type of question, but the expression of the traditional pre-relativistic point view. According to the latter, there are electrons, atoms, crystals, microscopes, people, buildings, planets, and galaxies. Objects or things, either simple or complex, exist and have definite properties at *each moment of time*. This view is *incompatible* with the special theory of relativity, as we will later see. Instead, we should consider collections of events and inquire into *the relations between them*. As Albert Einstein, in *"The Meaning of Relativity"*,[2] expressed it:

> ▷ **Einstein on Events**
>
> *It is neither the point in space, nor the instant in time, at which something happens that has physical reality, but only the event itself.*

We will return to this important point later in the book.

An idealized event is similar to a geometrical point: it occupies no volume and happens instantly. Realistic events, like the ones listed above, have finite duration and spatial extent; but both can be neglected in certain situations. Such idealization is analogous to how we treat planet Earth as a point in its motion around the sun, since the earth's size is negligible compared to the distance to the sun.

Once we adopt the event-based point of view, the similarity between geometry and physics becomes very strong. First of all, we can speak of the totality of all events or *event space*. Each point of this space is a separate event, of which we may ask physically interesting questions. For example, we may measure the separation in space (distance) or in time (duration) of one event from another; or we may measure how much energy, electric charge, mass, angular momentum was involved in the event, and so on.

---

[2] A. Einstein, *The Meaning of Relativity*, Lecture II.

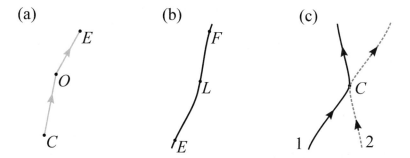

Fig. 3.1 Graphical representation of events and the relations between them: (**a**) The event C causes the event O, which causes the event E; (**b**) A continuous chain of events; (**c**) Two chains of events that "connect" at the event C. This picture may describe a collision of two particles

▷ Spacetime

In the special theory of relativity event space is called *spacetime*.

The basic characteristics of an event are *when* (space) and *where* (time) it occurs.

Special relativity shows that spacetime is not an artificial sum of space separately and time separately, but indeed a unity of events.

Two events in spacetime may be related in some sense. Figure 3.1a shows three events, such that the event O is *caused* by the event C, and in its turn causes the event E. The event E is thus the *effect* of the event O, and the event O is the effect of the event C.

A continuous chain of events, shown as a line in Fig. 3.1b, is an idealized representation of some series of events. Consider, for example, a light lamp that is first turned on (event E), and later turned off (event F). During the interval indicated by the events E and F the lamp is emitting light; each emission event represented by some point L on the line connecting the events E and F.

Lines of events can be used to illustrate physical interactions, such as a collision of two particles. Figure 3.1c shows the collision event C between particles 1 and 2. Many physical phenomena can be represented graphically in a similar way.

Events, graphically represented, are also called *world points*—a term introduced by Hermann Minkowski. As a renowned physicist Max Born put it:[3]

▷ Physics According to Max Born

Physics is the doctrine of the relations between such marked world points.

---

[3] Max Born, *Einstein's Theory of Relativity*, Dover, 1965, p. 333.

The laws of physics, expressing the relations between events, are quite analogous to the laws of geometry, expressing the relations between points and figures. In this sense, physics explores the *geometry of spacetime*. This point will be further elaborated after we get acquainted with spacetime diagrams.

3.2   Frame of Reference

Events can be detected or observed, and their properties, such as location and time of occurrence, mass, charge, energy, etc., can be measured. To observe an event we need a detector or an *observer*.

▷ **Observer**

Observer is an experimental setup used to detect events and measure their properties.

An observer does not have to be human. Humans, as a matter of fact, are not particularly good at making accurate observations. Automated systems, performing measurements and data analysis, are now widely used both on earth and in space. It is better to think of such observers when studying the special theory of relativity.

To refer to an event we need to specify when and where it happened. To this end, an observer must have a *coordinate system* and a *clock*. Time t and three spatial coordinates (x, y, z) can then be assigned to an event (see Fig. 3.2). Four numbers (t, x, y, z), uniquely specifying an event, serve as coordinates in spacetime. Spacetime is four-dimensional.

▷ **Frame of Reference**

An *observer* capable of measuring *time t* and *location* (x, y, z) of an event is called a *frame of reference*.

In geometry we are free to choose any coordinate system from an infinite number of equally valid coordinate systems. The choice of a coordinate system is determined by a problem at hand and does not affect the conclusions about geometrical laws. Do we have the same freedom of choice for a frame of reference? Are all observers equally valid in their observations of events? The principle of relativity and its experimental confirmation gives an answer to this. But before we take a further look into this issue, we need to examine how times and locations of events are measured.

Fig. 3.2 Key components of a frame of reference: a coordinate system and a clock. A reference frame is used to label events with time and space coordinates (t, x, y, z)

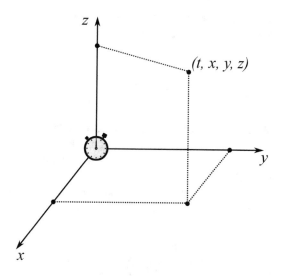

3.3   Measuring Time

Frame of reference acts as a spacetime coordinate system, since it allows identification of events by assigning them unique combinations of time and position coordinates. We first take a closer look at how a time of an event can be measured.

3.3.1   What Is Time?

In his book *"Relativity and Common Sense,"*[4] a cosmologist Sir Hermann Bondi expressed the following view on time:

▷ **Time According to Bondi**

Time is that which is measured by a clock. This is a sound way of looking at things. A quantity like time, or any other physical measurement, does not exist in a completely abstract way. We find no sense in talking about something unless we specify how we measure it. It is the definition by the method of measuring a quantity that is the one sure way of avoiding talking nonsense about this kind of thing.

But what exactly does a clock measure? How does it happen that different clocks measure the same time? To answer these questions we need to examine what a clock

---

[4] H. Bondi, *Relativity and Common Sense*, Anchor Books, New York, 1964.

is. We will review several types of clocks, examine their principles of operation, and see what they have in common.

3.3.2 Mechanical Clock

Mechanical clock uses some kind of a mechanism producing a periodic motion. Grandfather's clock and an alarm clock with a manual winding are good examples of mechanical clocks. The physical principles of their operation are illustrated in Fig. 3.3.

The mechanism of an actual clock can be quite complicated, but three main components are always present:

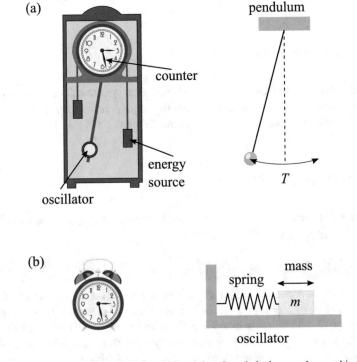

Fig. 3.3 Two types of mechanical clocks: (**a**) Pendulum-based clocks use a heavy object attached to a long arm, swinging back and forth under the force of gravity; (**b**) Spring-based clocks use a combination of a spring and a mass. Inside an alarm clock there is a spiral-like spring that is compressed and decompressed in a periodic motion

> ▷ **Main Components of a Clock**

1. Oscillator.
2. Energy source.
3. Counter.

An *oscillator* is the part of a clock that makes the clock tick *uniformly*—the time between any two ticks is the same. A good oscillator, or a *frequency standard*, is essential for a good clock.

A *counter* keeps track of the number of ticks (oscillations) the oscillator performed. It registers the advancement of time.

An *energy source* keeps the oscillator going by providing energy, which is constantly being lost due to friction or some other process.

A pendulum-based clock, like grandfather's clock, uses a pendulum as an oscillator, weights as an energy source, and a dial as the counter.

A mass attached to a spring, or a massive spring alone, can perform periodic oscillatory motion. A specially designed spring is used in a wind-up alarm clock as an oscillator. The manual winding-up provides the energy, and the dial keeps track of the number of ticks.

3.3.3   Electronic Clocks

Figure 3.4a shows a quartz wristwatch and its main oscillating component—a U-shaped quartz crystal resembling a tuning fork. When a quartz crystal is strained, a voltage drop appears on its surface. This is a *piezoelectric effect*, discovered in 1880 by Pierre and Jacques Curie. The effect works in the opposite direction too: When a voltage is applied to a quartz crystal, the latter becomes strained and changes its shape slightly.

A quartz crystal, cut out in a shape of a tuning fork, vibrates with a certain fixed *frequency*—number of vibrations per unit of time. As the quartz crystal vibrates, the strain changes periodically, creating a periodic voltage which can be measured via metal contacts (see Fig. 3.4a and c). The battery in a wristwatch keeps the quartz crystal vibrating by periodically "kicking" it through a voltage supplied by a special circuitry. Thus, the quartz clocks are not purely electronic since they rely on mechanical vibrations of a crystal.

Figure 3.4b shows a purely electronic device—the most popular integrated circuit, called "555" timer. It can be made using about 40 electronic components, such as transistors, resistors, and diodes, all compactly manufactured in a silicon chip. The output signal of such a device, similar to the output of a quartz crystal, is a periodically varying voltage applied to one of its pins. This periodic voltage can

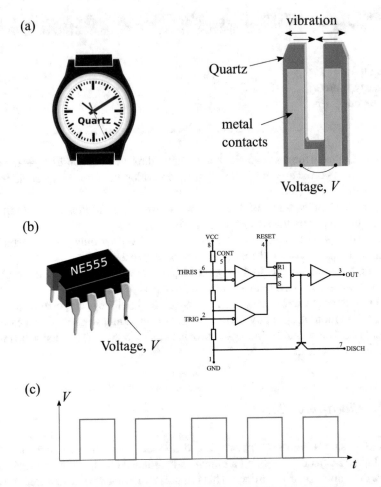

Fig. 3.4 Two example of clocks using electronic effects. (**a**) Quartz clocks, where a piece of crystal is used to produce oscillations and create electronic signal (voltage). (**b**) The most popular integrated circuit: an electronic "555" timer, invented in 1971 by Hans Camenzind. The schematic on the right illustrates that the timer is made of several electronic components. (**c**) An electronic clock produces a periodic voltage that changes between low and high values. A dedicated circuit counts each transition and converts the number of oscillations into a time reported on the clock's display

be detected by a counter displaying the elapsed time. The schematic for the "555" timer, shown in Fig. 3.4b, is given just to illustrate its purely electronic nature.

A quartz clock uses a quartz crystal as an oscillator; the battery acts as an energy source, and a dial as a counter. The oscillator in the "555" timer is a special electronic circuit, made of several electronic components and hidden inside the chip. The energy source can be an external battery; the counter can also be an external circuit.

Electronic clocks are generally more accurate than mechanical clocks. In addition, they are stable, less sensitive to environmental factors, such as temperature changes and shaking. Last, but not the least, they are cheaper.

3.3.4 Atomic Clocks

To achieve higher accuracy, stability, and reproducibility (identical operation of independently constructed clocks) scientists developed atomic clocks. There are different variants of such clocks, all with sophisticated designs and numerous technical nuances. Below we consider the physical principles of a cesium clock—an atomic clock based on the atoms of cesium.

Every atom has a nucleus, where the positive charge and most of the atom's mass is concentrated, and electrons which are distributed around the nucleus. Figure 3.5a shows (schematically) a nucleus and a single electron. Cesium has 55 electrons, all but one forming pairs; the unpaired electron is essential for the operation of cesium-based atomic clocks.

Both the nucleus and the unpaired electron of a cesium atom behave like tiny magnets. They can, like any pair of magnets, have one of two distinct mutual orientations: either pointing in the same direction with their "north poles", or pointing in the opposite directions, as illustrated in Fig. 3.5a. Similar to a pair of big magnets, the tiny magnets of the nucleus and the unpaired electron prefer to point in the opposite directions; this corresponds to the middle of Fig. 3.5a and denoted as a filled circle with small anti-parallel arrows: ($\uparrow\downarrow$). A certain amount of energy E is required to change the mutual orientation of the "atomic magnets" from anti-parallel to parallel, the latter configuration is denoted with open circle and two parallel arrows inside: ($\uparrow\uparrow$). The energy can come from an electromagnetic wave of a certain frequency f.

To flip the relative orientation of the electronic and nuclear "magnets", the frequency of the electromagnetic wave must be fixed with a very high accuracy. For cesium atoms the required frequency equals

$$f_0 = 9\,192\,631\,770 \text{ Hz}$$

or oscillations per second.[5] The electromagnetic field of this frequency f_0 can transfer a *quantum* of energy

$$E_0 = hf_0$$

to the cesium atom and flip the relative orientation of the magnets. The number h, called Planck constant, is the fundamental physical constant discovered by Max

---

[5] A CPU of a modern computer has a clock running at about one third of this frequency.

(a)

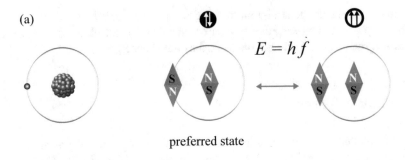

preferred state

(b)

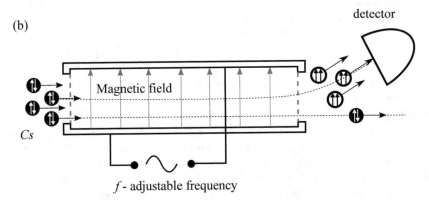

Fig. 3.5 Principle of operation of an atomic clock: (**a**) An atom with an unpaired electron that interacts with the magnetic field of the nucleus; (**b**) A tube with an oscillating magnetic field that "flips" the atomic magnets if the frequency f of the magnetic field matches the stable and precise atomic frequency f_0. "Flipped" atoms can be detected at the output end of the tube

Planck in 1900.[6] The energy E_0 is relatively small, equal to the kinetic energy of a single cesium atom moving with the speed 7 meters per second.

The physical principle of the operation of a cesium-based atomic clock is schematically illustrated in Fig. 3.5b. A beam of cesium atoms enters a tube with oscillating magnetic field. Initially, all atoms are prepared in a state with "magnets" pointing in the opposite directions (black circles on the left part of the figure). If the frequency f of the magnetic field matches the atomic frequency f_0, the reorientation of the nuclear and electronic "magnets" happens and the atoms leave the tube in a different state (shown as open circles on the right side of the figure). A special device at the output of the tube deflects "flipped" atoms to a detector, where they are

---

[6] $h = 6.62607004 \times 10^{-34}$ J s.

Table 3.1 Characteristic frequencies of different atoms that can be used as frequency standards

Clock	Frequency (Hz)
Cesium	9,192,631,770
Rubidium	6,834,682,610.904
Hydrogen	1,420,405,751.786
Strontium	429,228,004,229,873.4

counted. If the frequency of the magnetic field f does not match f_0, no atoms will flip the orientation and the detector will have no output, but when the frequency f matches the atomic frequency f_0, the signal from the detector becomes the strongest.

The cesium atoms act as filters or "selectors" of a precise frequency of the magnetic field. The frequency of the magnetic field f can be adjusted until it matches f_0, using the signal from the detector as a feedback. Once the magnetic field frequency f is *locked* onto atomic frequency f_0, a dedicated electronics counts the oscillations of the magnetic field, producing "ticks" of the atomic clock.

Cesium is one of several elements that are used in atomic clocks. Different atoms have different characteristic frequencies; some of them are listed in the Table 3.1. The choice of a particular atom is dictated by a compromise between the clock's performance and ease of use for specific application. What works for a terrestrial research laboratory may be very unpractical for a spacecraft.

? Problem 3.1*

Imagine two observers far apart and at rest relative to each other. Both observers have identically built clocks and can communicate with each other by exchanging signals. Devise a procedure which can be used by the observers to synchronize their clocks. In other words, find a way to make sure that these observers will have the same time on their clocks.

3.3.5   Random Clocks*

A perfect clock runs uniformly, with equal durations between any two successive "ticks". All real clocks, including all the clocks considered above, have some imperfections; *no two clocks will "tick" absolutely in sync.*

Interestingly, even a "clock" that produces completely *random* "ticks" can be of some use. To construct a random clock we can use a naturally occurring random process of radioactive decay. To illustrate a principle, we can take neutrons—subatomic particles present in every atomic nucleus except hydrogen. A single neutron is *unstable*, it eventually decays into a positively charged proton, a negatively charged electron, and an elementary particle called electronic anti-neutrino. The decay can be written as a formula:

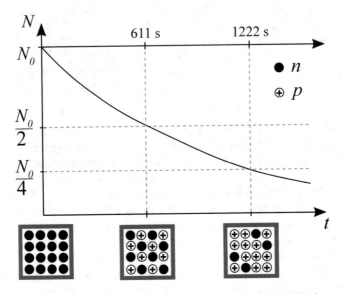

Fig. 3.6 The number of unstable particles decreases with time. The half-life of neutrons is 611 seconds, meaning that (on average) after 611 seconds only half of the initial number of neutrons remains

$$n \rightarrow p + e + \bar{\nu}_e.$$

The charged product particles fly away and can be detected, thus providing a way to count the number of decayed neutrons.

As illustrated in Fig. 3.6, if we start with N_0 neutrons, then after 611 seconds about half of them will decay. After the next 611 seconds, about half of the remaining neutrons will decay, leaving only $N_0/4$ neutrons. It is said that the *half-life* of neutrons is 611 seconds.

The decay process is unpredictable: There is no certainty that given two neutrons only one will remain after 611 seconds. However, if we repeat the experiment with two neutrons many times, in approximately half of the trials a single neutron will remain. The more repetitions we do, the closer to half of the cases we get. This is why we say that *on average* half of neutrons decay during the half-life.

How can this be useful for building a clock? If we can detect and count each decay, then we would have a counter that "ticks" and marks the passage of time. It would be possible to say that "the event E happened M ticks ago," and everyone using the *same* random clock would know what moment of time we are referring to. This approach is not fundamentally different from using the number of full moons to indicate when an event happened. The essential difference is that the "ticks" of a random clock happen unpredictably and non-uniformly, making its use inconvenient, but not impossible.

In 1949 an American physical chemist Willard Frank Libby developed a method of measuring the "age" of archeological objects made of organic materials. The method is based on the decay of radioactive carbon into nitrogen, electron, and anti-neutrino:

$$^{14}_{6}C \rightarrow \ ^{14}_{7}N + e + \bar{\nu}_e.$$

The radioactive carbon is continuously produced in the upper atmosphere; it then propagates lower, ends up in carbon dioxide and eventually is consumed by plants in the process of photosynthesis. When a plant dies, the supply of new radioactive carbons stops, and the already present atoms decay with half life of 5730 years. The decays can be detected and from their *rate* (number of decays per unit of time) it is possible to tell how long ago the plant died.

3.3.6 Absolute Time and Universal Time*

Today, there are more man-made clocks than there are people on the planet. There are also man-made clocks in outer space. The clocks are based on different physical phenomena: mechanical, electronic, atomic, as reviewed above, and others. Despite a great progress in the definition and measurements of time, the task of the determination of exact time and its global distribution is still a nuanced and difficult problem.

Most of the clocks are *not synchronized*: they show different time and run at different rates. It is an acceptable situation for most people, as long as their clocks are tolerably accurate and periodically adjusted. Science and technology demand much higher accuracy and stability of separate clocks, as well as effective synchronization of clocks in different locations.

Relatively recently in the history of science and technology, the definition of the unit of time was still based on the duration of a day. The day is split into 24 hours, an hour is divided into 60 minutes, and a minute is divided into 60 seconds. A second, therefore, is 1/86,400 of a day.

A moderately accurate mechanical clock will show that two consecutive days will have different duration; the difference can be as high as 30 seconds. Therefore, a simple definition of a unit of time based on the rotation of the Earth is not sufficiently accurate. Fortunately, the length of a day averaged over a year, called *mean solar day*, is much more constant.

It is known that even the mean solar day is not constant—it is getting longer as the rotation of the Earth is slowing down. Every 1000 years the mean solar day grows by 17 milliseconds (17/1000 of a second; for comparison, a human eye blink takes about 100 milliseconds). Surprisingly, using very careful observations astronomers were able to detect such a tiny effect before the invention of accurate atomic clocks. What could they use as a clock, that is as precise and stable as atomic oscillations?

To find the slowing down of a given "clock", we can compare it to another clock, known to be more stable and accurate. The discrepancy between the number of "ticks" the two clocks produce will grow with time, and thus reveal the difference in their stability. This is a simple and direct way, but unfortunately not available for astronomers. Instead, they rely on accurate observations of the positions of various celestial objects.

Positions of comets, planets and their satellites can be predicted if we precisely know three things: (1) their initial positions; (2) their initial velocities; (3) laws of their motion given by some equations (e.g., Newton's law of gravity and Newton's second law of motion). Once the equations are solved for specified *initial conditions*, the predicted position will depend only on time for which the prediction is made. The important question is this: *The time of which clock should one use for this calculation?* The best available time was based on mean solar day. Gradually, the observations of Moon revealed the difference between its actual position (including its positions in the distant past, as reported by historical records) and the position predicted using the time based on the mean solar day. The Moon appeared to be "accelerating," and it was understood that this apparent acceleration is due to the slowing down of Earth's rotation.

> **Newtonian Time**

To match the astronomical observations with predictions based on Newtonian laws of motion, we must correct the time of whatever imperfect clock we use in practice to the "true" or *Newtonian time*.

Isaac Newton expressed his view on time as follows:[7]

" *Absolute, true, and mathematical time, of itself, and from its own nature flows equably without regard to anything external, and by another name is called duration; relative, apparent, and common time, is some sensible and external (whether accurate or unequable) measure of duration by means of motion, which is commonly used instead of true time.*"

For several years, before the invention of atomic clock, Newtonian time, deduced from a number of astronomical observations, was used for the standard of time. Such time is called *ephemeris time*. Since 1967, however, the unit of time is based on cesium clock.

Now should we just build the most accurate clock possible and use it to tell what time it is for everyone, everywhere? This approach is impractical. If we use it, then the passage of the sun through the meridian (noon) will be occurring not at exact 12 o'clock, but a bit later, due to the slowing down of the Earth's rotation. When the

---

[7] Isaac Newton, *Principia Mathematica*, Scholium on Absolute Space and Time, 1687.

clock runs uncorrected, the noon will eventually move to 1 minute past 12, then to 2 minutes, and so on indefinitely, creating unnecessary confusion. To prevent this from happening, the difference between the precise time of an atomic clock and the time based on the rotation of the Earth is deliberately kept within 1 second. This is the idea behind the Coordinated Universal Time, used by whole planet and International Space Station.

By the international agreement, the responsibility to provide universal time is given to *International Bureau of Weights and Measures* (BIMP). Various laboratories around the world measure time using their own local atomic clocks. Then, the readings from various laboratories are sent to BIMP, where the data is processed and the official time for the world is calculated.

3.3.7 "Time Itself"

Various clocks used in practice rely on different physical phenomena, such as mechanical, electronic, atomic, etc. Different clocks run at different rates and possess different levels of randomness and imperfections. No two clocks, placed next to each other, will "tick" perfectly in sync. And yet, we tend to think that every clock measures, however imperfectly, the same thing—"time itself".

In the view of Isaac Newton, all our clocks are imperfect attempts to measure the *absolute time* (ideal, true time) as accurately as possible. This view is incompatible with the special theory of relativity and has been criticized by many scientists. Henri Poincare, in *"The Measure of Time"*,[8] wrote:

▷ **Poincare on Time**

In other words, there is not one way of measuring time more true than another; that which is generally adopted is only more convenient. Of two watches, we have no right to say that the one goes true, the other wrong; we can only say that it is advantageous to conform to the indications of the first.

Thus, there is *no* universal "flow of time", which different clocks can "hook up" to and measure the same true and absolute time. There is nothing to "flow": There are no particles of time or anything that one could "put in a jar", figuratively speaking. *There is no "time itself".*

One should not get an impression that time is eliminated in special relativity. Time still is, as Sir Hermann Bondi put it, *"that which is measured by a clock."* Later

---

[8] H. Poincare, *The Measure of Time*, The Foundations of Science (The Value of Science), New York: Science Press, 1913, pp. 222–234.

we will learn how special relativity affects our understanding of time measurements, as well as such time-related notions as simultaneity, past, present, and future.

3.4 Measuring Distance

Distance measurement is a familiar task. To find, for example, the distance between two trees, we may take a straight stick and determine how many times it fits between the trees. If exactly N sticks fit, then our answer will be N meters, provided the stick is 1 meter long. This procedure is simple, intuitive, and direct. It relies on a very important physical concept—*rigid body*, used as a *measuring rod*. Rulers and meter sticks are examples of measuring rods used in many practical situations.

But how can we measure the distance between atoms, or the distance between the Earth and the Moon, or the distance between two bodies in motion? Different *methods* or *measurement procedures* must be used in these cases. Let us consider some of them.

3.4.1 Interatomic Distance

An accurate measurement of distances between atoms in a crystal can be done using electromagnetic radiation with very small wavelength, as illustrated in Fig. 3.7.

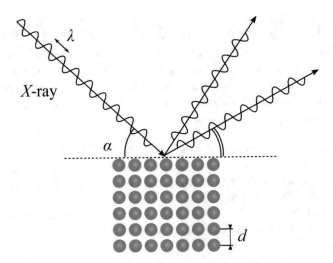

Fig. 3.7 Measuring interatomic distance using the method of X-ray scattering. An incident X-ray with the wavelength λ is scattered in distinct directions that depend on the distance d between atoms

Fig. 3.8 Measuring distance using the method of triangulation. Applying the laws of geometry, it is possible to find the distance to a remote object, like a tree on the other side of the river, without crossing the river

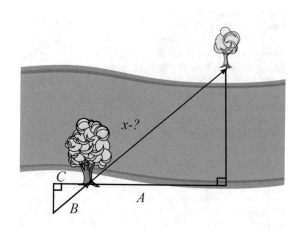

X-ray is a type of electromagnetic radiation; it is more energetic than ultraviolet light and has much shorter wavelength λ. When an X-ray is incident on a crystal, it is scattered in *several* distinct directions, determined by the angle if incidence α, the wavelength λ, and the distance between the atoms d. Measuring the angles of the incident and scattered light, and knowing the wavelength of the X-ray, it is possible to *calculate* the distance between atoms.

It is important to realize that this method relies on the accurate knowledge of the X-ray wavelength λ, which has to be measured separately.

3.4.2   Triangulated Distances

Suppose we want to find the distance between two trees, separated by a river that can not be crossed, as shown in Fig. 3.8. With the help of geometry we do not have to cross the river, we can do all the required measurements on just one side!Looking at the right triangles in Fig. 3.8, we see that the distance x equals

$$x = A\frac{B}{C},$$

where the distances A, B, and C are measured on the same side of the river. This method is called *triangulation*, it is used in astronomy to measure distances to moderately remote stars (see Fig. 3.9 for the illustration of the parallax method).

The fact that *we relied on the theorems of planar geometry is very important*. If the trees were growing on two opposite sides of a very curvy hill or valley, our use of geometry would change. Similarly, when astronomers are using the method of parallax, they assume that the light from the stars propagates along straight lines of three dimensional Euclidean space. It is assumed that the electromagnetic radiation

Fig. 3.9 Given the distance between the Earth and the Sun (1 astronomical unit) and the angles (α and β) at which the same star (S) is observed at two locations half a year apart, we can deduce the distances, a and b, to the remote star. For a very distant star the distances a and b are almost the same

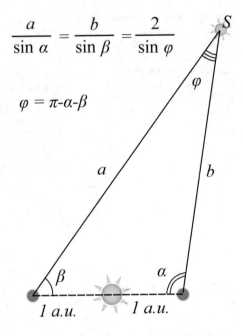

$$\frac{a}{\sin \alpha} = \frac{b}{\sin \beta} = \frac{2}{\sin \varphi}$$

$$\varphi = \pi - \alpha - \beta$$

(light from the stars) propagates along the straight line from the star to the Earth—the shortest path, when measured with a regular ruler.

3.4.3 RADAR Method

The word radar stands for **RA**dio **D**etection **A**nd **R**anging. As the name implies, the method is used to find distances to objects using radio signals. It involves sending a radio signal to an object to be reflected, then detecting the reflected signal, and measuring the round-trip time of the signal. A radio signal is a type of electromagnetic radiation; it is significantly less energetic than infrared light and has much longer wavelength, ranging from a centimeter to a meter.

To better understand how the radar method works, consider the source of a signal, the detector of the signal, and the clock, as illustrated in Fig. 3.10. The source, the detector, and the clock are all in the same place.

A short pulse signal is emitted at time t_1. The reflected signal is received at a later time t_2. The time it takes for the signal to travel from the emitter to the reflecting object and back is[9]

$$\Delta t = t_2 - t_1.$$

---

[9] See Appendix A.2 for the comments on using Δ notation.

Fig. 3.10 Measuring distances to remote objects using an electromagnetic signal. An electromagnetic pulse can be sent to a remote object. The pulse is reflected and received back. Based on the emission time t_1, detection time t_2, and the speed of the signal, the distance to the object can be calculated, as well as the time of reflection t_R

Traveling back and forth the signal covers the distance

$$L = 2d,$$

where d is the distance between the emitter and the object *at the moment of reflection*. The last qualification is important, because the object reflecting the signal may be moving relative to the emitter/detector. If the speed of the signal v is known, we can find the total traveled distance

$$L = v\Delta t.$$

The distance from the emitter to the object at the moment of reflection is then

$$d = L/2 = v\Delta t/2 = v\frac{t_2 - t_1}{2}.$$

Note that since the detection always happens after the emission, $t_2 > t_1$, the quantity d is always positive. Additional information, like the orientation of the radar, is needed to assign a position (positive or negative) on the x axis. In the future, we assume such information is available, and we can properly determine the time t and location x for every event.

The radar method also gives the time t_R when the event of reflection happens. To travel from the source to the object, the signal spends half of the round-trip time Δt. If the signal was emitted at t_1, then the reflection happens at

$$t_R = t_1 + \Delta t/2 = \frac{t_2 + t_1}{2}.$$

? Problem 3.2

Suppose that the time of reflection $t_R = t$ and the position $d = x$ of the object at that time are known. Express the time of emission t_1 and detection t_2 in terms of t and x.

▷ LIDAR

The signal used in the radar method does not have to be a radio signal. It can be any type of electromagnetic signal, including the visible light. This variant of radar method is called *Lidar*, for light radar. A green laser light, used in Apache Point Observatory (New Mexico, USA), allowed the measurement of the distance to a special "mirror" on the Moon with a millimeter accuracy.

3.5   Spacetime Diagrams

The location and the time of an event can be measured using any method. The result is a set of four numbers—(t, x, y, z)—representing the temporal and the spatial coordinates of the event, measured in the specified reference frame. Other events will get different coordinates; various relationships between events can be studied by looking at their coordinates.

In a geometrical approach, events are represented graphically as points in special diagrams, called *spacetime diagrams*. Spacetime diagrams often provide a simple graphical way to analyze relationships between events. We do not have to use spacetime diagrams, sticking to words, algebra, and imagination instead. However, when mastered, spacetime diagrams can be very helpful.

3.5.1   2D Spacetime Diagrams

Spacetime diagrams look simplest when the events under study all happen on the same axis. Consider, for instance, an object moving along a straight line, as shown

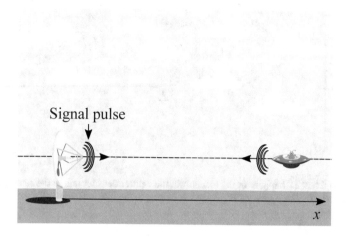

Fig. 3.11 An object moving along a straight line can be tracked using the radar method. An event happening to the object receives a pair of coordinates (t, x)

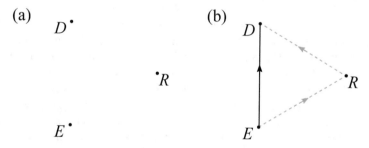

Fig. 3.12 Spacetime diagrams illustrating the radar method. (**a**) Three key events of the radar method: E—emission of the signal; R—reflection; D—detection. (**b**) Additional elements of the spacetime diagram that show the connection between the key events: propagation of the signal between the emission and reflection, and between the reflection and the detection (directed dashed-line segments)

in Fig. 3.11. Imagine that the radar method is used to find the objects' positions at various moments of time.

The location of the objects using the radar method involves at least three separate events: E—emission of a signal, R—reflection of the signal from the object, and D—detection of the reflected signal, when it arrives back at the source. These events can be represented as three points in spacetime, arranged in a certain way, as illustrated in Fig. 3.12a. The placement of the points E, R, and D reflects the fact that the events happen at different times and, possibly, in different places.

In addition to the events E, R, and D, we may indicate many other events. For example, we may connect the event E with the event R using a line showing all events where the propagating signal was present. We could put an arrow on this

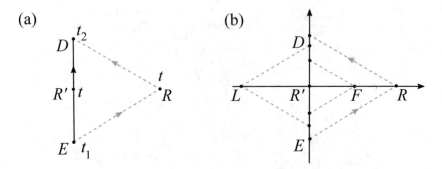

Fig. 3.13 (**a**) Event R' happens at the location of the emitter/detector and at the same time as the remote event R. (**b**) The events L, R', F, and R belong to the line indicating all events happening at the same time

line, pointing out that the chain of events unfolds in the direction from E to R; see Fig. 3.12a. A similar directed line segment can be drawn from R to D.

Additionally, we can connect the events E and D with a straight line to highlight all events that happen to the source/detector between the emission and detection of the radio signals. This way we get a *figure in spacetime*—the triangle $\triangle ERD$.

Between the events E and D there must be an event, denoted as R' in Fig. 3.13a, that happens at the same time as the event R. As the analysis of the radar method showed, this moment must be half-way between the events E and D: $ER' = R'D$.

Figure 3.13b shows two more events *simultaneous* with R—the events F and L. The event F can be viewed as a reflection of a radar signal from an object closer to the source than the object at R. The times of emission and detection for the event F might be different from corresponding times for the event R, but the half-way time can still be the same. Similarly, the event L can be simultaneous with R, as long as the half-way time between the times of the emission and detection for the event L coincides with R'. In fact, the "horizontal" line through the events $LR'FR$ indicates the events that happen at the same time. It is common sense to view all events that happen at the same time as *the world at a given moment of time*. This view, as we will later discover, is incompatible with the special theory of relativity.

If the source and the detector are in the same location, we can place the origin of the coordinate system at the detector and assign times and positions to different events using the radar method. In Fig. 3.14a, the temporal and the spatial coordinates for the events E, R, and D, involved in the radar method, are indicated, together with the coordinates of two events simultaneous with R—the events F and R'.

The emission event E has the *spacetime coordinates* $(t_1, 0)$, the detection event D has the coordinates $(t_2, 0)$. Given the speed of the signal v, the spacetime coordinates of the reflection event R are

$$R \rightarrow (t, x) = \left(\frac{t_1 + t_2}{2}, v\frac{t_2 - t_1}{2} \right).$$

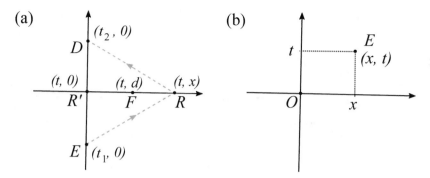

Fig. 3.14 (a) Events in spacetime diagrams can be assigned coordinates (t, x), indicating when and where the events happen. (b) The axes x and t of spacetime diagram, analogous to the Cartesian axes x and y, but with different units for the horizontal and the vertical directions

The event F happens at the same time as R, but at the location closer to the origin ($d < x$).

In Fig. 3.14a, the line through the events E, R', and D indicates all events that happen at the origin, $x = 0$, at various moments of time. The line through the events R', F, and R indicates all events that happen at same time, $t = 0$, at different locations. We must keep in mind that the measurements of time and position of an event is always done relative to a specific coordinate axes and a clock; in other words, *the measurements are done relative to a specific reference frame*.

Given a reference frame, any event can be graphically represented using a set of axes similar to the Cartesian coordinates; see Fig. 3.14b. Such a diagram must not be confused with Cartesian coordinate system because the meaning of the axes is different. The horizontal axis represents various positions x at the same time of the clock, while the vertical axis represents the same position $x = 0$ at various times of the clock.

▷ **Spacetime Is Non-Euclidean**

An important difference between Euclidean plane and planar spacetime diagrams is the following: In Euclidean plane all direction are equivalent, whereas in spacetime diagrams two different directions can have very different meaning.

Figure 3.15 shows two spacetime diagrams with certain events indicated using straight lines. In Fig. 3.15a the vertical line corresponds to all events that occur to the object at the origin, while horizontal lines indicate simultaneous events for three different moments of time: now ($t = 0$), earlier ($t_- < 0$), and later ($t_+ > 0$). Such horizontal lines represent one dimensional world at various moments of time.

The vertical lines in Fig. 3.15b indicate events that happen at different moments of time at fixed locations: At the origin, $x = 0$; on the left of the origin, $x_- < 0$; on the right of the origin, $x_+ > 0$. For example, the leftmost vertical line may

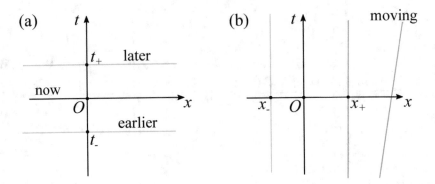

Fig. 3.15 (a) Events that happen at the same time are indicated in spacetime diagram as horizontal lines. (b) Events that happen in the same place at various times are indicated as vertical lines. Lines with a slope correspond to moving objects

correspond to all events happening to an object at rest relative to the radar. The rightmost line in Fig. 3.15b has a visible slope, it indicates events happening at various locations at various moments of time. It might describe an object moving away from the radar.

3.5.2 3D Spacetime Diagrams

To fully specify an event one needs all four spacetime coordinates (t, x, y, z). When only two spacetime coordinates (t, x) matter, spacetime diagrams become two-dimensional; such diagrams were introduced in the previous section. When three spacetime coordinates (t, x, y) are sufficient, the spacetime diagrams become more complicated.

Figure 3.16 shows a three-dimensional spacetime diagram. The vertical line t, like before, indicates all events happening at the origin at various times; the axis x corresponds to all events that happen at the same time, $t = 0$, and have the same spacetime coordinate $y = 0$; the axis y corresponds to all events that happen at the same time, $t = 0$, and have the same spacetime coordinate $x = 0$.

All simultaneous events form a plane, three of which are shown in Fig. 3.16. The plane marked as "now" contains all events with the temporal coordinate $t = 0$; the plane marked as "earlier" contains all events with the same time $t_- < 0$; finally, the plane marked as "later" is made up from all events with the same time $t_+ > 0$.

As an illustration, each plane contains an event, marked as a black dot, corresponding to a location of some object, as it moves in two spatial dimensions. The line through all three points indicates all events where the object could be detected. For an object moving in circles, like a planet around a star, the spacetime diagram is shown in Fig. 3.17.

Three dimensional spacetime diagrams are not used often.

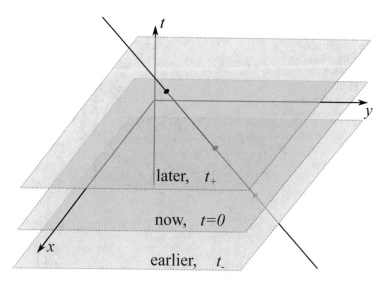

Fig. 3.16 Three dimensional spacetime diagram. The xy planes represent two-dimensional "universe" at fixed moments of time: now, earlier, and later

Fig. 3.17 Spacetime diagram with events corresponding to an object moving in a circular orbit in the xy plane

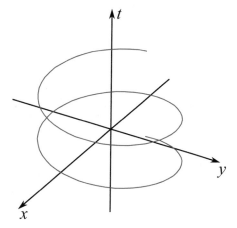

3.5.3 4D Spacetime Diagrams

When all four spacetime coordinates are required for the analysis of events, it becomes challenging to clearly represent events in spacetime diagrams. Visualizing events in four dimensions is, generally, not easy, but still possible.

Let us consider a specific example of two objects moving in three spatial dimensions, as shown in Fig. 3.18. Although the paths of the objects cross, the objects may visit the same point at different moments of time and thus avoid the collision. This is easily seen using a spacetime "triptych" shown in Fig. 3.19.

Fig. 3.18 Paths of two
different objects moving in
three dimensional space.
Although the lines intersect at
the point P, the objects do not
necessarily collide, because
they can visit the same point
at different moments of time

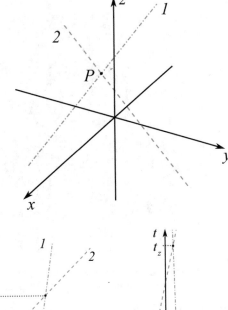

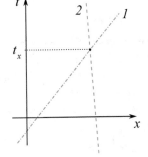
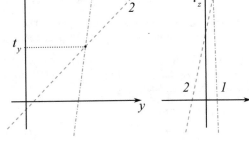

Fig. 3.19 Combination of three spacetime diagrams helps represent events from four dimensional
spacetime. In this example, two objects clearly do not collide, because their spatial coordinates
(x, y, z) coincide at different moments of time $t_x \neq t_y \neq t_z$

The method of representing four dimensional spacetime using three separate
spacetime diagrams is similar to the way three dimensional objects are represented
using three projections: front, top, and side views.

? Problem 3.3

Use spacetime diagram to show two pulses sent from the radar at times t_1 and
t_2. Show how these signals can be reflected from two different objects, located at
difference distances d_1 and d_2 from the radar, but **at the same time** t, according to
the frame of reference of the radar.

? Problem 3.4

Draw spacetime diagrams showing a radar locating an object at two different moments of time. That is, show two events, E_1 and E_2, of the emission of signals, two events, R_1 and R_2, of the reflection of signals, and two events, D_1 and D_2, of the detection of signals. Consider three cases (i.e., draw three spacetime diagrams):

(1) The object being located did not move between the events R_1 and R_2.
(2) The object was approaching the radar between R_1 and R_2.
(3) The object was receding from the radar between R_1 and R_2.

How are the time intervals between (E_1, E_2) and (D_1, D_2) are related for all three cases (i.e., is the time between the emitted signals longer, shorter, or the same compared to the time between the received signals)? How are the time intervals between (E_1, E_2) and (R_1, R_2) are related for all three cases?

? Problem 3.5

Two observers, A and B, are in relative motion. According to A, the observer B is initially far away and is approaching with a constant speed v. The observer B is carrying a clock that has a flashing light illuminating the dial every second, as measured by this clock. Essentially, every second the observer B sends an image of its current time to A. Draw the spacetime diagram of this process and explain what A will see (visually perceive). Consider the situation when the observer B is approaching, passing by, and then receding from the observer A.

Chapter Highlights

- Event is the basic concept in the physical picture of the world. Event-based view is the first major step towards understanding the special theory of relativity.
- In the special theory of relativity, spacetime is the totality of all events.
- The study of relationships between events is the goal of physics. This makes physics similar to geometry, which studies the relationships between abstract points in abstract spaces.
- Events are observed and measured by *observers*—experimental setups that detect events. Observers are not necessarily human.
- An observer with a clock and a method to measure the position of an event constitutes *a frame of reference*.
- When and where a given event happens are the basic pieces of physical information that are measured about the event. In general, there are three spatial coordinates and one temporal coordinate used to locate an event, making spacetime four-dimensional.

- Time is a physical quantity measured by a clock. Clocks of different designs and operational principles measure similar time *by design*. Physicist design their clocks in such a way that the laws of physics take on the simplest form.
- Distance between a pair of events can be measured using a clock and a signal. In the radar method electromagnetic signals are used.
- Events and relationships between them can be graphically represented using spacetime diagrams. The latter can be helpful, but they are not mandatory.

References

1. Einstein, A. (1979). *Autobiographical notes: A centennial edition*. Open Court, LaSalle and Chicago.
2. Einstein, A. (1970). *The meaning of relativity*, Lecture II. Princeton University Press. pp 16–17.
3. Born, M. (1965). *Einstein's theory of relativity* (p. 333). Dover.
4. Bondi, H. (1964). *Relativity and common sense*. Anchor Books.
5. Newton, I. (1995). *The principia, scholium on absolute space and time*. Prometheus. Translated by Andrew Motte.
6. Poincare, H. (1913). The measure of time. In *The foundations of science (The value of science)* (pp. 222–234). Science Press.

Chapter 4
Galilean Relativity

Every observed change of place is caused by a motion of either the observed object or the observer or, of course, by an unequal displacement of each. For when things move with equal speed in the same direction, the motion is not perceived, as between the observed object and the observer.

N. Copernicus, *The Revolutions of the Heavenly Spheres*, 1543.

Abstract In this chapter we discuss the views on space, time, and motion from the standpoint of Galileo and Newton. The distinction between absolute and relative quantities in mechanics is made. Inertial reference frames are defined. Transformation of time and space coordinates of events between different inertial reference frames is derived. Finally, the ideas are applied to solve several problems.

4.1 Relative and Absolute in Mechanics

We encountered in planar geometry quantities of two kinds: relative and invariant (or absolute). Cartesian coordinates of a point are the quantities of the first kind — they have different values in different Cartesian coordinate systems. Distance between a pair of points, or an angle between a pair of lines, are examples of invariant quantities. Although coordinates are useful, they are not essential in the formulation of geometrical laws. An analogous approach is used in physics: a physical quantity can be either relative or absolute. Absolute quantities are important because they are independent of the "point of view" — the specific reference frame. We now explore the main mechanical quantities: position (distance), time interval, velocity, and acceleration; the goal is to establish their relative or absolute character.

© The Author(s), under exclusive license to Springer Nature Switzerland AG 2022
Y. Deshko, *Special Relativity*, Undergraduate Lecture Notes in Physics,
https://doi.org/10.1007/978-3-030-91142-3_4

4.1.1 Relativity of Motion

By the time you finish reading this sentence, your position in space will change by several hundred of thousands of meters. The Earth is moving around the Sun with the speed of 30,000 meters per second; the solar system is moving with the speed of 200,000 meters per second around the center of the Milky Way; the latter is moving with the speed of 600,000 meters per second among the extra-galactic objects. Despite such high speeds, these motions are not detected by our senses.

For some types of motion near the surface of the Earth, it appears impossible to tell whether the motion is actually happening. Many early philosophers and scientists realized this fact, but we owe it to Galileo Galilei the formulation of a *principle*, stating that identical experiments, performed in identical laboratories, moving relative to each other, will produce identical results, *"so long as the motion is uniform and not fluctuating this way and that"*.[1] In other words, *rectilinear uniform motion* of an observer is not detectable by experiments.

In his 1632 book, *"Dialogue Concerning the Two Chief World Systems"*, Galileo described an imagined laboratory placed *"in the main cabin below decks on some large ship"* . An observer in the laboratory studies various physical effects, such as the free fall of water droplets, the motion of fish in a bowl, projectile motion and others. Galileo concludes: *"You will discover not the least change in all the effects named, nor could you tell from any of them whether the ship was moving or standing still."* Although Galileo speaks about the same observer in different states of motion (rest vs. rectilinear uniform motion), it is an equally valid statement when applied to two different observers: one at rest, relative to the ground, and the other moving rectilinearly and uniformly relative to the first observer.

In Galileo's time a ship smoothly gliding across the ocean was the best example of a rectilinear uniform motion. Today we may talk about a spaceship moving with constant velocity, far from influences of stars, planets, or other massive objects. Without looking outside and seeing, for example, an approaching planet, it will be impossible to determine whether the spaceship is moving.

In the mechanics of Galileo and Newton the uniform motion is *relative*: Only the change of the position of an object relative to another object can be detected. There is no physical experiment that would tell whether the laboratory is uniformly moving *itself*, *absolutely*, or, in the words of Newton, *" without regard to anything external"*.

---

[1] *Dialogue Concerning the Two Chief World Systems*, transl. by S. Drake, University of California Press, 1953, pp. 186–187.

4.1.2 Invariance of Durations and Lengths

The mechanics of Galileo and Newton uses the following assumptions about the measurements of time and distance between any two events:

> **Assumptions of Galileo-Newtonian Mechanics**

- *Time interval between any two events is the same for all observers.*
- *Length of any object is the same for all observers.*

These assumptions are confirmed by everyday experiences and form a part of "common sense" view of the world. We will use these assumptions in this chapter to derive the relationship between time and position of any event, as measured by a pair of observers in relative motion. Later, we will analyze these Galileo-Newtonian assumptions from the standpoint of the special theory of relativity.

4.1.3 Galilean Transformation

When two different observers measure the position and the time of the same event, they generally get different results. Figure 4.1 shows two observers, S and S', separated by a distance d. From the figure it is clear that if the observer S measures the position of an event E as x_E, then the observer S' must measure the position as

$$x'_E = x_E - d.$$

If the observers use identical clocks and synchronize them for convenience, then the times of the event E must be the same:

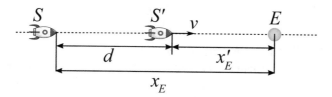

Fig. 4.1 The relation between the spatial coordinates x_E and x'_E of the same event E. *Note:* In this figure the point E is *not* a point of spacetime, but only its projection onto the x axis; in other words, this is not a spacetime diagram, this is a space diagram

$$t'_E = t_E.$$

The relative distance d between the observers may be changing in time. Suppose the observer S is tracking the position of the origin O' of the reference frame S'. Then from Fig. 4.1 follows that

$$x_{O'} = d,$$

and the velocity of S' relative to S is

$$v = \frac{\Delta x_{O'}}{\Delta t} = \frac{\Delta d}{\Delta t}.$$

The principles of Galilean relativity and the special relativity are applicable *only to the observers moving rectilinearly and uniformly*; therefore, the direction and the magnitude of the relative velocity v is constant. If the initial distance between the observers S and S' is d_0, then at the moment of time t it equals to

$$d = d_0 + vt.$$

To keep formulas simple, we will assume that at $t = 0$ the observers S and S' are in the same location on the x axis and, therefore, $d_0 = 0$. The relation between the spatial coordinates of the event E then takes the form

$$x'_E = x_E - vt_E.$$

We can drop the subscript E, since the relations between the time and position are true for *any* event. The result is a set of equations, called *Galilean transformation of spacetime coordinates*:

$$t' = t, \tag{4.1}$$

$$x' = x - vt, \tag{4.2}$$

where

$$v = \frac{\Delta x_{O'}}{\Delta t}$$

is the velocity of the reference frame S' relative to S.

The inverse transformation is easily found:

$$t = t',$$
$$x = x' + vt'.$$

The formula for x looks similar to Eq. (4.2), with the sign of the velocity flipped.

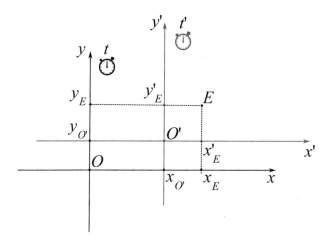

Fig. 4.2 The relation between spatial coordinates of the same event, as measured by two observers in relative motion in two dimensions

4.1.4   Galilean Transformation in 2D and 3D*

In general, the observer S' can be moving relative to S in an arbitrary direction. The transformation formulas (4.1) and (4.2) were derived for the special case of one-dimensional motion along the x axis. A more general case can be analyzed in a similar way.

Figure 4.2 shows the coordinate axes of the reference frames S and S' in the case of two-dimensional relative motion. For an arbitrary event E, the spatial coordinates, as measured by S and S', are related in the following way

$$x'_E = x_E - x_{O'} = x_E - v_x t_E,$$

$$y'_E = y_E - y_{O'} = y_E - v_y t_E,$$

where

$$v_x = \frac{\Delta x_{O'}}{\Delta t}, \quad v_y = \frac{\Delta y_{O'}}{\Delta t}$$

are the velocities of the reference frame S' relative to S along the axes x and y, respectively.

If the relative motion involves the third axis z, then an additional relation exists:

$$z'_E = z_E - z_{O'} = z_E - v_z t_E,$$

where

$$v_z = \frac{\Delta z_{O'}}{\Delta t}.$$

For an arbitrary event, the Galilean transformation is written as follows:

$$t' = t,$$
$$x' = x - v_x t,$$
$$y' = y - v_y t,$$
$$z' = z - v_z t.$$

Often a vector form of these equations is used:

$$\mathbf{r}' = \mathbf{r} - \mathbf{v}t, \quad t' = t, \tag{4.3}$$

where $\mathbf{r}$ (or $\mathbf{r}'$) is the vector connecting the origin of the coordinate system of S (or S') and the location of the event; $\mathbf{v}$ is the vector of relative velocity.

The Galilean transformation for two- and three-dimensional motion have similar form; they express the same idea as the transformation for one-dimensional motion. To understand the theory of relativity, either Galilean or special, it is sufficient to consider the motion only along a single axis.

4.1.5 Relativity of Velocity

Consider a situation when two observers, S and S', are both tracking the position of the same object; both observers use the radar method (Fig. 4.3). The observer S measures that at the moment of time t_1 the object has the position x_1, then, at the later time t_2, the position of the object changes to x_2. The observer S calculates the velocity of this object, assuming it was moving uniformly, as

$$u = \frac{\Delta x}{\Delta t} = \frac{x_2 - x_1}{t_2 - t_1}. \tag{4.4}$$

The value of u can be positive or negative, depending on the direction of motion of the object relative to S.

The observer S' calculates the velocity of the same object similarly:

$$u' = \frac{\Delta x'}{\Delta t'} = \frac{x'_2 - x'_1}{t'_2 - t'_1}.$$

The spacetime coordinates (t'_1, x'_1) and (t'_2, x'_2) can found from the Galilean transformation, given by the relations (4.1, 4.2):

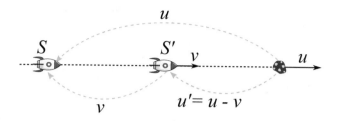

Fig. 4.3 Two observers in relative motion measure the velocity of an object. The velocities of the object relative to each observer are related as $u' = u - v$

$$x_1' = x_1 - vt_1, \quad t_1' = t_1,$$

and

$$x_2' = x_2 - vt_2, \quad t_2' = t_2.$$

Plugging these equations into the expression for u', we find

$$u' = \frac{x_2 - x_1 - v(t_2 - t_1)}{t_2 - t_1} = \frac{\Delta x - v\Delta t}{\Delta t} = u - v.$$

Given the velocity of the object in the reference frame S', we find that relative to S it will be moving with the velocity

$$u = v + u'.$$

It is important to remember: v and u can be either positive or negative, depending on the direction of the relative motion of the reference frames and the object.

We showed that velocity is a physical quantity defined *relative to the reference frame*: the velocity of the same object is different in different frames of references; in other words, it is not an absolute or invariant quantity.

4.1.6 Invariance of Acceleration

Suppose the observer S' finds that the speed of an object at the moment of time t_1' is u_1' and later, at the time t_2', it changes to u_2'. If the speed of the object was changing uniformly with time, the observer S' will find the acceleration:

$$a' = \frac{\Delta u'}{\Delta t'} = \frac{u_2' - u_1'}{t_2' - t_1'}.$$

At the same moments of time $t_1' = t_1$ and $t_2' = t_2$, the observer S will measure the speed of the object to be

$$u_1 = v + u_1', \quad u_2 = v + u_2',$$

respectively. The acceleration, according to S, is

$$a = \frac{\Delta u}{\Delta t} = \frac{u_2 - u_1}{t_2 - t_1} = \frac{u_2' - u_1'}{t_2' - t_1'} = a'.$$

This is an important result, stating that *the acceleration of an object is the same for all observers in rectilinear uniform relative motion*. If the object is not accelerating relative to S $(a = 0)$, and the observer S' is not accelerating relative to S $(v = \text{const})$, then the object can not be accelerating relative to S' $(a' = 0)$. The object will be either at rest or moving with constant speed relative to the reference frames S, S', or any other reference frame moving rectilinearly and uniformly relative to S.

In Galileo-Newtonian mechanics, acceleration is an absolute (invariant) quantity.

4.1.7 Length of a Moving Object

Everyday experience tells us that lengths of rigid objects should be the same, whether they are at rest or moving rectilinearly and uniformly. Let us see how this agrees with Galilean transformation.

Consider two observers, S and S', in relative motion with the velocity v. The observer S' has a stick at rest in its reference frame; the stick, therefore, is moving with the velocity v relative to the observer S, as shown in Fig. 4.4.

The length of the stick can be found from the measurements of the positions of its left and right edges:

$$L = x_R - x_L,$$

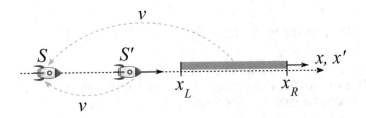

Fig. 4.4 Two observers in relative motion measure the length of a stick by locating its left and right edge. For a moving stick both edges must be located at the same time

for the observer S, and

$$L' = x'_R - x'_L,$$

for the observer S'.

The observer S must be careful to measure both x_L and x_R *at the same time*, otherwise the stick will shift during the measurement. (For the observer S' the measurements do not have to be simultaneous, since the stick is stationary in its reference frame.)

Thus, we have two measurement events:

1. Event R, with the spacetime coordinates (t, x_R).
2. Event L, with the spacetime coordinates (t, x_L).

For the observer S' these events happen at (t'_R, x'_R) and (t'_L, x'_L). From Galilean transformation follows that

$$t'_R = t = t'_L,$$

$$x'_R = x_R - vt,$$

$$x'_L = x_L - vt.$$

Finally,

$$L' = x'_L - x'_R = x_L - x_R = L.$$

Both observers will measure the same length; in other words: The length of a rigid object is an invariant (absolute) quantity in Galileo-Newtonian mechanics.

4.1.8 Invariance of Laws

The principle of Galilean relativity can be formulated as follows:

▷ **Principle of Galilean Relativity**

The laws of mechanics have the same expression in all reference frames that move rectilinearly and uniformly.

This formulation is neither unique nor perfect, but we can learn a lot from clarifying it.

Laws of Mechanics

In 1687 Sir Issac Newton published a work, *"Mathematical Principles of Natural Philosophy"*, containing the foundations of what is now known as Newtonian

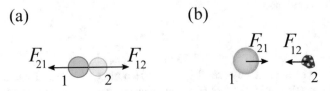

Fig. 4.5 Newton's third law: Equality of action and reaction ($F_{12} = -F_{21}$). F_{12} is the force the body 1 exerts upon 2, F_{21} is the force the body 2 exerts upon 1. (**a**) Bodies touch each other during an interaction (e.g., collision); (**b**) bodies interact at a distance (e.g., a star attracts a comet)

mechanics. Newton proposed several laws that successfully explained and predicted a wide range of motions: From objects moving near the earth's surface, to planets moving through space. Let's review these laws, in the reverse order.

③ *Newton's third law* states that interacting bodies exert on each other *forces* of equal magnitudes but of opposite directions, as illustrated in Fig. 4.5.

② *Newton's second law* states that under the action of a given force, a body undergoes change of momentum proportional to the force. In modern notation this is expressed using the famous formula

$$F = ma.$$

To see the connection with momentum, recall that in the Newtonian mechanics the latter is given by $p = mv$; for the constant mass we have

$$\Delta p = m \Delta v = ma \Delta t,$$

where we used the definition of acceleration. From the last expression follows

$$F = \frac{\Delta p}{\Delta t} \quad \text{or} \quad \Delta p = F \Delta t.$$

① *Newton's first law* states that when no force is applied to an object (or all forces balance each other) the latter either remains at rest or moves rectilinearly and uniformly.

Three laws of Newton, combined with the expressions for various forces (elastic spring force, force of universal gravitational attraction, constant force, etc.) can describe how objects move through space with time.

Invariance of Expression

To understand the phrase *"The laws of mechanics have the same expression in all reference frames"* we will consider a specific example, illustrated in Fig. 4.6.

Let's assume that an observer S finds that two bodies, with the masses m_1 and m_2, are moving according to the equations

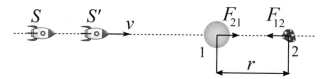

Fig. 4.6 Two observers, moving uniformly and rectilinearly relative to each other, will have the same equations for the motion of bodies under mutual gravitational attraction

$$m_1 a_1 = G \frac{m_1 m_2}{r^2}, \tag{4.5}$$

$$m_2 a_2 = G \frac{m_1 m_2}{r^2}, \tag{4.6}$$

where a_1 (or a_2) is the acceleration of the body 1 (or 2), as measured by the observer S; r is the distance between the bodies; G is the gravitational constant. These equations follow from Newton's second law of motion combined with Newton's law of universal gravitational attraction.

The observer S' will obtain exactly the same equations for the following reasons:

1. Acceleration of any body is *invariant* for any two observers in relative rectilinear uniform motion.
2. Lengths and distances are *invariant* as well.
3. Gravitational constant is a fundamental quantity—the same for all observers.
4. In Galileo-Newtonian mechanics the mass of any object is the same in all reference frames.

Thus, the left and right sides of Eqs. (4.5, 4.6) are the same for both observers S and S'.

> More generally, physicists speak about *form-invariance* of laws. What is meant by this is the following: Although the numerical values of the left and right sides of an equation, expressing the law of physics, change between the reference frames, the *equality* of both sides remains true, and the meaning of terms on each side remains the same.

Inertial Frames of Reference

The point discussed next may appear subtle, but it is very important. It deals with the *rectilinear and uniform motion* that appears in Newton's first law of mechanics and in the relative motion of any two observers considered so far. We want to clarify the concept of inertial observers, and why they are important in Galileo-Newtonian mechanics, as well as in the special theory of relativity.

(a)
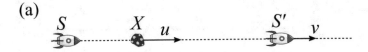

(b)
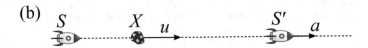

(c)
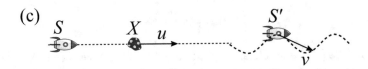

Fig. 4.7 Different variants of a relative motion of two observers: (**a**) Uniform and rectilinear; (**b**) rectilinear but accelerated/nonuniform; (**c**) non-rectilinear motion with constant speed v. Only in (**a**) both observers can be inertial—both find the law of inertia true

Consider an observer S, tracking an object X, far removed from other bodies; see Fig. 4.7. Imagine that the object is also far enough from the observer, so we can neglect their mutual gravitational attraction. Thus, no detectable forces are applied to the object X; the latter must either be at rest relative to S ($u = 0$), or be moving with constant speed along a straight line ($u = const$), according to Newton's first law.

(a) Now consider another observer S', moving relative to S and also sufficiently far from the object X to produce any effect on the latter. If the observer S' is moving uniformly and rectilinearly relative to S, as shown in Fig. 4.7a, then the velocity of the object X relative to S' will be

$$u' = u - v = const.$$

Relative to S', the object either remains at rest ($u = v$) or moves with constant velocity in a straight line ($u \neq v$). If the object X were moving with an acceleration a relative to S, then it would have the same acceleration relative to S', as we have shown before.

(b) If the observer S' is moving rectilinearly but *non-uniformly*, e.g., with constant acceleration a, as illustrated in Fig. 4.7b, then the object X can not maintain the same velocity relative to S'. Newton's first law fails for S', since the object X does not move uniformly, despite the absence of external forces.

(c) Finally, if the observer S' is moving along a curved path (even with constant magnitude of the velocity), as in the Fig. 4.7c, then relative to S' the object X

can not move rectilinearly. Again, Newton's first law fails for S', since the object X does not move rectilinearly, despite the absence of external forces.

The conclusion is the following: If the laws of mechanics are true in some reference frame S, then they will be equally true and have the same expression in any other reference frame S', if the latter is moving rectilinearly and uniformly relative to S. The laws of mechanics will not be true for a reference frame moving non-uniformly or non-rectilinearly (or both) relative to S. This is why we limit ourselves only to reference frames where the laws of Newtonian mechanics apply—*inertial reference frames*.

***

To conclude the discussion of inertial reference frames, let's take another look at Newton's first law: *When no force is applied to an object (or all forces balance each other) the object remains at rest or moves rectilinearly and uniformly*. This law can not be true for an arbitrary observer; we showed above that a rectilinear and uniform motion in one reference frame may become non-rectilinear or non-uniform in another. There must be a reference frame implied in Newton's first law of mechanics, and it is relative to this reference frame the motion of a force-free object is rectilinear and uniform. What is this reference frame? Does it exist? Is it unique?

Newtonian mechanics works with high degree of accuracy for motions near the surface of the earth and for the motions of celestial objects. Therefore, the reference frames associated with the earth or the solar system are good examples of *inertial reference frames—reference frames where Newtonian mechanics works* (with good enough accuracy). As we showed, if the laws of Newtonian mechanics hold in one reference frame, they hold in any other, as long as the reference frames are in relative rectilinear and uniform motion. There are infinitely many such reference frames, all qualify as inertial reference frames.

4.2   Absolute Space and Time*

The reference frames associated with the earth or sun are not perfectly inertial, since both are not moving rectilinearly. Newton's laws of mechanics hold true in these reference frames only approximately. For Isaac Newton, the inertial reference frames associated with the earth, sun, or any other body, were imperfect approximations to an ideal reference frame not bound to physical objects. In his 1687 book *"Mathematical Principles of Natural Philosophy"*, Newton postulated that there must exist the ultimate inertial reference frame—*absolute space* and *absolute time*—relative to which laws of Newtonian mechanics hold absolutely accurately.

If a reference frame is at rest, or moving rectilinearly and uniformly, relative to Newton's absolute space, it is an inertial reference frame; Fig. 4.8 shows two such reference frames, S and S'. It is impossible to detect rectilinear uniform motion of a

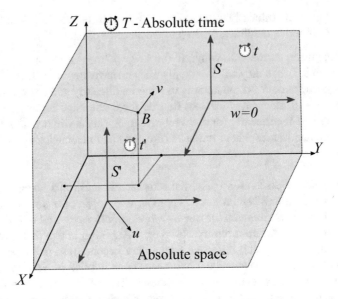

Fig. 4.8 Absolute space and time in Newton's picture of the world provide the ultimate reference frame. In this case it makes sense to ask "is an object really/absolutely moving"? Here, the reference frame S is at rest relative to the absolute space, while the reference frame S' is moving relative to the absolute space with the velocity **u**. Observers in both reference frames will discover the same laws of mechanics

reference frame, as expressed in the principle of Galilean relativity. Yet, in Newton's view, it is meaningful to speak of an *absolute motion* of an object—motion relative to the absolute space.

The idea of the absolute space and the absolute reference frame was analyzed by many scientists for a very long time. Physicists tried to find physical experiments that would allow to detect the rectilinear uniform motion relative to this absolute reference frame, unsuccessfully. The special theory of relativity shed a new light on the concept of absolute time and its relation to the absolute space. We will learn about it in later chapters. To get a glimpse of how modern physics understands space and time, we will turn to a contemporary physicist, who specializes in these questions.

Space and Time Disappear

Italian physicist Carlo Rovelli, an expert in loop quantum gravity, masterfully summarized modern views on space and time in his paper *"The Disappearance of Space and Time"*.[2]

---

[2] Carlo, Rovelli (2006). *The disappearance of Space and Time*. In Dennis Geert Bernardus Johan Dieks (ed.), *The Ontology of Spacetime*. Elsevier.

First, Rovelli points out that *"Special Relativity is little more than a minor variation of the Newtonian conceptualization of spacetime"* and that *"we must focus on General Relativity if we want to hold a view of space and time compatible with what we have understood so far about the natural world"*. Einstein's theory of general relativity is well outside of the scope of this book; suffices to mention that this is the theory of gravity, quite different from Newtonian theory. Similar to the theory of electromagnetism, general relativity uses the idea of a *physical field*—an extended entity, in contrast to material points which are localized in space.

Physical fields can affect motion of material points; for example, electric or magnetic fields affect the motion of an electron. Moreover, different physical fields can affect each other, or be affected by material particles. Thus, physical fields are not "frozen" or fixed things, they can change and evolve while interacting. They are *dynamical object*. Gravitational field is such an object, analogous to the electromagnetic field, but more complex.

Rovelli next writes: *"Einstein's discovery is that Newtonian space and time and the gravitational field are the same entity"*. Recall that the Newtonian absolute space was like an infinite container for all physical objects. In addition, absolute space determined the motion of particles free from forces—*inertial motion*, rectilinear and uniform motion relative to the absolute space and time. General relativity modifies this view; as Carlo Rovelli puts it: *"The clean way of expressing Einstein's discovery is to say that there are no space and time: there are only dynamical objects. The world is made by dynamical fields. These do not live in, or on, spacetime: they form and exhaust reality. One of these fields is the gravitational field."*

As we see, the idea of the eternal and rigid "container for things" or "stage for events" is gone. Only dynamical fields remain, which can interact and affect each other. Gravitational field is now responsible for determining what it means for an object to be in an inertial motion, and the inertial motion is, in general, no longer rectilinear and uniform. Furthermore, in general case, gravitational field is not constant and fixed, but can be affected by energy and motion of other dynamical fields (e.g., electromagnetic field) and objects.

As Albert Einstein put it:[3] *"Physical objects are not in space, but these objects are spatially extended. In this way the concept "empty space" loses its meaning."*

4.3 Application of Galilean Relativity

According to the principle of Galilean relativity, all inertial reference frames are equally valid for applying the laws of mechanics. This gives a freedom to choose any inertial reference frame to analyze a problem. The situation is analogous to the freedom of choice of a coordinate system in geometry. The Exercise 2.2 showed that properly chosen coordinate system makes problem simpler. Similarly, properly

---

[3] A. Einstein, *Relativity: The Special and General Theories*, Note to the Fifteenth Edition.

Fig. 4.9 A heavy elevator is moving upwards and elastically collides with a ball. The time between successive collisions of the ball with the elevator's floor can be found in any inertial reference frame

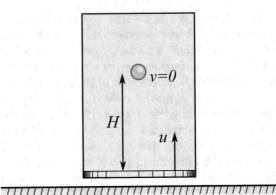

chosen reference frame may make a problem in mechanics simpler, as the following example illustrates.

Consider a heavy elevator moving upwards with constant speed u relative to the ground. At a certain moment, an elastic ball is released, with zero initial speed, from the height H above the elevator's floor, as shown in Fig. 4.9. The ball begins falling down, due to the earth's gravitational pull, and eventually hits the floor. After the elastic collision it bounces off with the same speed (as required by the elastic collision), goes up and then down, to collide with the floor again and again. We need to find the time between successive collisions of the ball and the elevator's floor.

The problem is formulated in the reference frame of the ground—an inertial reference frame. Let us consider this problem from the reference frame of the elevator. The elevator is moving with constant speed relative to the ground and is also an inertial reference frame.

In the reference frame of the elevator, the ground is moving down with the constant speed u, and the ball starts its fall from the height H above the floor with the initial speed u, as shown in Fig. 4.10a. The gravitational pull will accelerate the ball, so that right before it collides with the floor the ball will have speed $w > u$; see Fig. 4.10b. This speed can be found from the conservation of mechanical energy:

$$\frac{mu^2}{2} + mgH = \frac{mw^2}{2},$$

from which follows that

$$w = \sqrt{u^2 + 2gH}.$$

Immediately after the elastic collision, the ball will start moving upwards with the speed w. As the ball is ascending, its speed is decreasing until, after time t_{asc}, it becomes zero. Since the only force acting on the ball is due to gravity, the acceleration of the ball is

Fig. 4.10 In the reference frame of the elevator the ball is falling down with the initial speed $v' = u$. It eventually collides elastically with the floor, bounces off, goes up, stops, and then falls down to collide again. The process continues indefinitely

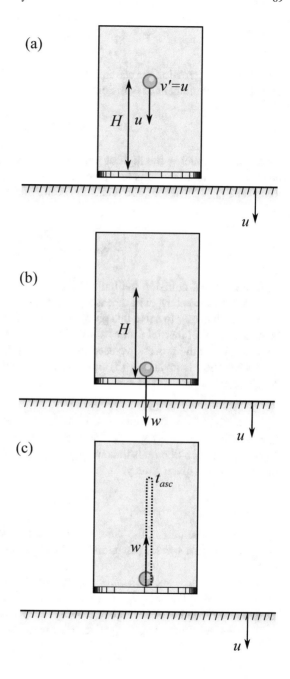

$$\frac{w}{t_{asc}} = a = g.$$

From the last expression we can find the time required to reach the highest point of ascent from the moment of collision:

$$t_{asc} = \frac{w}{g}.$$

After the ball reaches the highest point it will start falling down to the floor. The downward motion is a complete reverse of the upward motion, hence is takes the same time t_{asc}. Therefore, the total time between the first and the second collision is

$$T = 2t_{asc} = 2\frac{\sqrt{u^2 + 2gH}}{g}.$$

After the second collision the ball will start moving up with the speed w so the process described above will repeat again and again.

It is much harder to solve this problem in the reference frame of the ground. The equivalence of all inertial frames allowed us to choose a different point of view on the problem. In the process we used the following facts: (1) velocities are relative; (2) acceleration is invariant; (3) time between events is invariant; (4) the law of conservation of mechanical energy is invariant.

***

As another example, consider a problem illustrated in Fig. 4.11. An observer in one of the numerous galaxies discovers that all other galaxies are apparently receding from her. She finds that every galaxy has only radial speed proportional to the distance to a given galaxy:

$$v = Hr,$$

where H is a coefficient, the same for all galaxies. Can the observer conclude that her galaxy is in a special place, and all other galaxies are moving away from this special location?

To answer this question, we can consider the situation in another inertial reference frame. For simplicity, we will focus on just three galaxies, denoted as O, O', and O'' in Fig. 4.12, and all moving in the same plane.

The observer in the galaxy O finds that the speed of the galaxy O' is

$$v = Hr_{OO'},$$

and the speed of the galaxy O'' is

$$u = Hr_{OO''}.$$

Fig. 4.11 An observer measures that the galaxies are moving away from her. The speed of every galaxy is proportional to the distance to that galaxy. The same picture is observed in any other galaxy

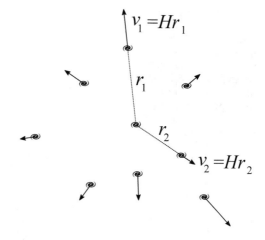

The reference frame S' of the galaxy O' is moving along both axes x and y with the velocities

$$v_x = v \cos \alpha = H r_{OO'} \cos \alpha = H x_{O'},$$

$$v_y = v \sin \alpha = H r_{OO'} \sin \alpha = H y_{O'}.$$

Similarly, the reference frame S'' of the galaxy O'' is moving relative to S with the velocities

$$u_x = u \cos \beta = H r_{OO''} \cos \beta = H x_{O''},$$

$$u_y = u \sin \beta = H r_{OO''} \sin \beta = H y_{O''}.$$

From the Galilean transformation for the motion along a single axis, we derived the relationship between the velocity u of an object as measured by two observers S and S' in relative motion with the speed v:

$$u'_x = u_x - v_x.$$

A similar relationship can be derived for the general case of relative motion along other axes. Namely, for the y axis we obtain

$$u'_y = u_y - v_y.$$

Using these formulas, we can find the velocity of the galaxy O'' relative to the galaxy O':

$$u'_x = H x_{O''} - H x_{O'} = H x'_{O''}, \tag{4.7}$$

Fig. 4.12 (**a**) An observer in the galaxy O finds other galaxies (e.g., O' and O'') to be moving radially away, with the speeds $v = Hr_{OO'}$ for O', and $v = Hr_{OO''}$ for O''. (**b**) If we switch into the reference frame of the galaxy O', all other galaxies will receive an additional "boost"—their velocities will get an additional term $-\mathbf{v}$

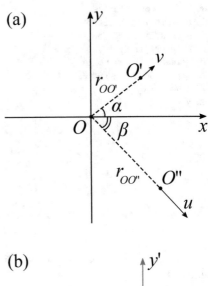

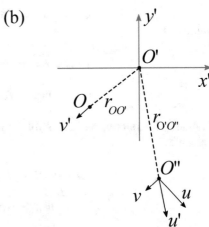

$$u'_y = Hy_{O''} - Hy_{O'} = Hy'_{O''}. \tag{4.8}$$

The magnitude of this velocity is

$$u' = H\sqrt{(x'_{O''})^2 + (y'_{O''})^2} = Hr_{O'O''},$$

it is proportional to the distance between the galaxies O' and O''. The motion is directed along the line connecting these galaxies, as follows from (4.7) and (4.8). The same answer can be obtained more elegantly using vectors. The conclusion is the following: There is no special galaxy that can be considered as the center; from the point of view of *any galaxy*, all other galaxies are receding with purely radial velocities, proportional to the distance to a given galaxy.

4.4 Practice

The following several problems should help get more comfortable with the principle of Galilean relativity. Although there is no universal algorithm for solving such problems, here are some helpful tips to keep in mind:

- First, identify all inertial reference frames in the problem.
- Use your best judgment and choose a reference frame. The more problems you solve, the better your judgment becomes.
- Re-calculate all relative quantities, such as velocities or position, if needed. Use velocity transformation formulas

$$u'_x = u_x - v_x,$$

$$u'_y = u_y - v_y.$$

- Use invariant quantities, such as accelerations, distances, masses, forces, and energies to set up equations for the solution.

? Problem 4.1

A perfectly elastic ball is moving towards a massive wall, while the wall is moving towards the ball. The speed of the ball relative to the ground is v; the speed of the wall relative to the ground is u; see Fig. 4.13. Find the speed of the ball relative to the ground after the elastic collision. To neglect the effects of gravity, assume that we are interested in the speeds right before and after the collision.

? Problem 4.2

A bucket is placed on a platform which is moving with constant speed v relative to the ground. The rain is falling vertically with the speed u, as shown in Fig. 4.14. How much should the bucket be tilted, so that the rain-drops do not hit the walls of the bucket.

Fig. 4.13 An elastic ball collides with a massive moving wall. Find the speed of the ball relative to the ground after the collision

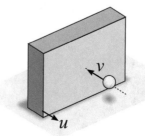

Fig. 4.14 A bucket on a
moving platform in the rain.
How much should the bucket
be tilted so that the rain-drops
do not hit the walls of the
bucket

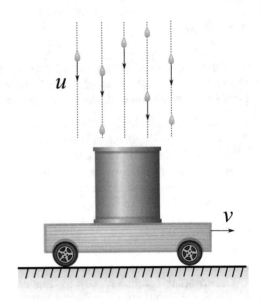

? Problem 4.3

When at rest, an unstable particle decays into two identical sub-particles, each
traveling with the speed u, as shown in Fig. 4.15a. Consider the case when the
unstable particle is moving with the speed v; see Fig. 4.15b. Find the maximum
possible angle that product-particles can have relative to the direction of the motion
of the original particle, i.e. relative to the dotted horizontal line in Fig. 4.15b.

? Problem 4.4

Two cars are moving with constant speeds in different directions, as shown in
Fig. 4.16. The initial distance between the cars is D. The direction of the motion
of the first car is given by the angle ϕ, measured relative to the line connecting the
cars at the start. The direction of the motion of the second car is given by the angle
θ. Find the minimal distance between the cars during their motion.

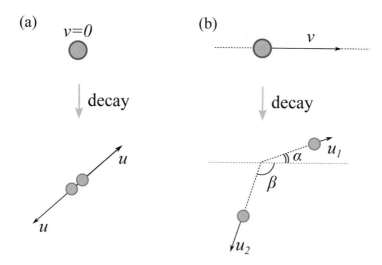

Fig. 4.15 (a) An unstable particle, if at rest, decays into two identical sub-particles, traveling in the opposite directions with equal speeds u. (b) A moving unstable particle decays into two identical sub-particles. Find the maximum possible angle between the trajectories of the original particle and any of its products

4.4.1 Additional Remarks

It is a good practice to consider the same problem in different reference frames, as it may lead to a simple picture of events and an easy solution. There is no guarantee, however, that this strategy will always work, as illustrated by the following exercise.

? Problem 4.5

Two massive walls are moving towards each other with speed v relative to the ground. A cockroach is running back and forth between the walls with constant speed $u > v$ (see Fig. 4.17). Having reached one wall, the cockroach turns around and runs towards the other wall, where it does the same. The initial distance between the walls is D; the cockroach starts at the left wall. Find the total distance traveled by the cockroach.

In this exercise there are at least three inertial reference frames: (1) the ground; (2) the left wall; (3) the right wall. The problem is stated in the reference frame of the ground. A reader is invited to consider the problem in all reference frames and see in which one it looks the simplest.

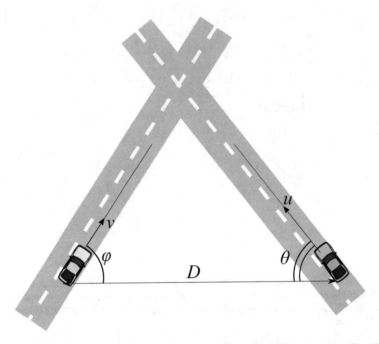

Fig. 4.16 Two cars, starting at the same time, are moving with different speeds. The distance between them is D, and the directions of their motions are specified the angles ϕ and θ. What is the minimum distance between the cars in the process of their motion?

Fig. 4.17 A cockroach is running back and forth between two closing walls. Find the total distance traveled by the cockroach

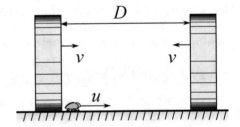

Chapter Highlights

- Experiments suggest that (absolute) uniform and rectilinear motion is undetectable.
- In Galileo-Newtonian physics time interval between two events is invariant (absolute)—the same for all observers. The distance between two events is also invariant.
- Velocity is a relative quantity. The velocity of a given object in different reference frames is found using simple "velocity addition rule."
- Acceleration is an absolute quantity in Galileo-Newtonian physics.

- Among the infinity of possible observers and reference frames, the *inertial reference frames* are singled out in the theory of relativity. Inertial reference frames are frames where the laws of Newtonian mechanics hold.
- The absolute space and absolute time, in the sense of Newton, are absent in modern physical theories.
- The freedom of choice of an inertial reference frame can be used to analyze a given problem from different points of view. Not infrequently, this leads to a simpler solution to the problem.

References

1. Copernicus, N. (1976). On the revolutions of the heavenly spheres. Translated by A. M. Duncan. London: David & Charles.
2. Galilei, G. (1953). *Dialogue concerning the two chief world systems* (pp. 186–187), transl. by S. Drake, University of California Press.
3. Carlo, R. (2006). The disappearance of space and time. In: D. G. B. J. Dieks (ed.), *The Ontology of Spacetime*. Elsevier.
4. Einstein, A. (2015). *Relativity: The special and general theories*, Note to the Fifteenth Edition.

Chapter 5
Special Relativity A

How it happened that I in particular discovered the relativity theory, it seemed to lie in the following circumstance. The normal adult never bothers his head about space-time problems. Everything there is to be thought about it, in his opinion, has already been done in early childhood. I, on the contrary, developed so slowly that I only began to wonder about space and time when I was already grown up. In consequence I probed deeper into the problem than an ordinary child would have done.

A. Einstein, *from "Albert Einstein: A Documentary Biography"* by Carl Seelig.

Abstract This is the central part of the book. Previous chapters introduced the principle of relativity, defined the notions of event, observer, and reference frame. Galilean transformation—the relationship between time and space coordinates of events for different inertial observers—implied that velocity, including the velocity of light, is different in different reference frames. This conclusion contradicts a large number of accurate experiments where the speed of light is measured under different conditions. The main goal of this chapter is to find a *proper way* to re-calculate the spacetime coordinates of events between inertial reference frames. *Proper way* means the way that agrees with the absolute nature of the speed of light. After we find the correct transformation, we will study their most important consequences.

5.1 Light Postulate

The following are firmly established experimental facts about the propagation of electromagnetic waves (including light):

© The Author(s), under exclusive license to Springer Nature Switzerland AG 2022
Y. Deshko, *Special Relativity*, Undergraduate Lecture Notes in Physics,
https://doi.org/10.1007/978-3-030-91142-3_5

▷ **Properties of Electromagnetic Signals**

- The speed of electromagnetic signals *in vacuum* does not depend on the direction of propagation.
- The speed of electromagnetic signals *in vacuum* is the same for all inertial observers.

These observations are used as the basic *postulates*—fundamental assumptions—in the special theory of relativity. We will use them to derive most of the results of the theory.

> By the speed of electromagnetic signals we mean the *round-trip* speed. There are nuances in talking about one-way speed of light, which we will not discuss in this book. Thus, the round-trip speed of the electromagnetic signal traveling north-south is the same as the round-trip speed for traveling east-west, and in any direction in between. As is commonly done, we will further assume that the speed of propagation of electromagnetic radiation in the opposite directions is also the same.

The propagation of electromagnetic waves is described by equation derived by Scottish physicist James Clerk Maxwell around 1860. Maxwell's equations, like equations of Newtonian mechanics, mathematically express a law of physics. According to the principle of relativity, the laws of physics, including the law of propagation of electromagnetic waves, must be the same for all inertial observers.

The *invariant* nature of the light propagation is incompatible with the Galilean transformation, since the latter requires *any* velocity be a *relative* quantity. Galilean transformation was derived using certain assumptions about the measurement of lengths and times which have to be modified.

> **But Why?**
> What makes electromagnetic waves so special that their speed is invariant? Actually, electromagnetic field is not the only physical field that propagates with the same speed for all observers. Gravitational field and the *strong field* (the field responsible for holding neutrons and protons together in stable nuclei, as well as for holding *quarks* inside protons and neutrons) also have this property. Thus, photons (particles of electromagnetic field),

(continued)

gravitons (particles of gravitational field), and gluons (particles of strong field) propagate with the same speed.

The existence of the invariant speed is connected with the existence of the *maximum speed of a signal* which can connect *cause and effect* events. The principle of relativity (the equivalence of all inertial reference frames) requires this maximum speed to be the same for all inertial observers.

This issue will be explored in more details, once we derive Lorentz transformation. See Appendix B.4 for further information.

5.2 Units

The concept of speed is usually introduced after the *procedures for the measurements* of length and time intervals are defined. If the distance is measured in meters and time in seconds, then the speed is measured in meters per second. A different approach is also used, as evidenced by the following statements:

> ▷ **Measuring Distances With Time**

- The post-office is 5 min of *walking* from home.
- The lake is 3 h of *driving* from the town.
- The stars are three light years apart (3 years of *light travel* apart).

In these examples the distances are defined using durations and some relevant speed (e.g., walking, driving, and the speed of light).

In Galileo-Newtonian mechanics *all* velocities are relative, while distances and time intervals are absolute. It seems more appropriate to use "distance and time" instead of "speed and time" to define other quantities. Nature however provides us with a special absolute standard of speed—the speed of light in vacuum.

The speed of light can be used as the scale for any other speed: the solar system travels at 0.00077 of the speed of light, the speed of the Earth due to its motion around the Sun is 0.0001 of the speed of light, and so on. Elementary particles in accelerators can achieve speeds higher than 0.99 of the speed of light, and electromagnetic waves propagate at the speed 1.

A light-year—a distance traveled by light in 1 year—is commonly used in astronomy. Similar to astronomers, we will be using the speed of light and duration to express distances. The distance between two objects will be measured using the radar method, and expressed as time required for a light signal to travel between

Table 5.1 Comparison of regular and c-based units. Starting with regular units we divide by the value of the speed of light in meters per second and get corresponding c-based value. To convert back into regular units—multiply by the same scale factor $\bar{c}$

Regular to c-based	c-based to regular
$c = 1(1)$	$\bar{c} = 3 \times 10^8$ (m/s)
$t\,(\text{s}) = \bar{t}$	$\bar{t}\,(\text{s}) = t$
$x\,(\text{s}) = \bar{x}/\bar{c}$	$\bar{x}\,(\text{m}) = x\bar{c}$
$v\,(1) = \bar{v}/\bar{c}$	$\bar{v}\,(\text{m/s}) = v\bar{c}$
$a\,(1\ /\text{s}) = \bar{a}/\bar{c}$	$\bar{a}\,(\text{m/s}^2) = a\bar{c}$

those objects: the Moon is 1.3 s away from us, the Sun is 500 s away, and the diameter of the Earth is 0.042 s.

In this new set of units—let's name it *c-based system*—distances have units of time. Time is left untouched, 1 s is still used as the unit of time. Velocities are measured in light-speeds and are expressed as fractions of the speed of light; they essentially become unitless.

? Problem 5.1

Determine the unit of acceleration in c-based system.

To avoid confusion, we will use two different notations for speed, time, distance, and acceleration: v, t, x, a will be used for the c-based units, and $\bar{v}, \bar{t}, \bar{x}$, and $\bar{a}$ for the usual, meter-based system of units. The comparison between the c-based and the regular units is given in the Table 5.1.

The benefits of using c-based system of units will become more evident as we progress.

For everyday phenomena the use of c-based units may seem very inconvenient. For example, the height of an average human is 0.0000000057 s, the acceleration due to gravity equals 0.000000033 $1/s$. Such tiny numbers result not only from the tremendous value for the speed of light (in meter-based units), but also from our tendency to use a "human" unit of time: An average human heart makes about 1 beat per second. In contrast, atomic phenomena have a very different time scale. The "heart" of a cesium atomic clock makes 9192,631,770 beats per second, making the time between two "ticks" equal to 1/9 192,631,770 of a second.

Physicist working with atomic processes prefer to use *nanosecond* as the unit of time. One second equals one billion nanoseconds. In one nanosecond a pulse of light will cover roughly 1/3 of a meter. The height of an average human is 5.7 nanoseconds, the acceleration due to gravity equals 32 $1/ns$.

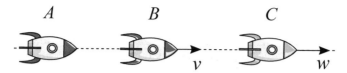

Fig. 5.1 Spaceships A, B, and C used as observers. They can freely move in an empty region of space, exchanging signals and collecting data

5.3 Observers

Earlier, when discussing the idea of observers in the Sect. 3.2, it was pointed out that any device capable of detecting signals, recording the results of measurements, and perhaps doing some calculations can be an observer. We will no, generally, use humans as observers.

Imagine three spaceships, identically designed and manufactured; all having long-lasting sources of power, fast and stable clocks, light emitters of various colors/wavelengths, detectors of electromagnetic radiation, cameras that save images in the infra-red, visible, and ultra-violet parts of electromagnetic spectrum; Fig. 5.1 shows three such spaceships, denoted as observers A, B, and C.

Astronomical observations indicate that the observable part of the Universe is quite empty. It is possible to find a volume of space large enough for the spaceships to travel for years at the speeds close to 1 without approaching any massive objects. Let us imagine a region of deep space of the diameter 10 light-years, free from stars, planets, asteroids, comets, large gas clouds, etc. It will be as empty as it gets, containing only three observers—A, B, and C—on their mission to discover the laws of relativity.

5.4 Minkowski Diagrams

It is possible to study relativity only using algebra. However, without taking anything away from formulas, we can benefit from graphical method for the analysis of events. Some effort is required to become comfortable with spacetime diagrams, especially after we switch to the c-based units.

Spacetime diagrams drawn using the c-based units are called *Minkowski diagrams*, after the German mathematician Hermann Minkowski. They differ from spacetime diagrams considered previously in two ways. First, the axes in Minkowski diagrams have the same units; in our case both have the units of time.

Second, in a diagram drawn for *any reference frame* the lines indicating the propagation of light-pulses (or other electromagnetic signals) are always represented by straight lines going diagonally. For example, the light-lines starting at the origin of a reference frame are represented in all Minkowski diagrams as straight lines described by the equations

$$x = ct = t \text{ or } x = -ct = -t.$$

The first equation corresponds to the light traveling in the positive direction of the x axis with speed $c = 1$; the second equation corresponds to the light traveling in the negative direction of the x axis.

5.4.1 Minkowski Terminology

When Minkowski introduced geometrical approach to relativity in 1908, he used several terms that became standard. Firstly, a point in a spacetime diagram (an event) is called a *worldpoint*. Secondly, the sum total of all worldpoints is called the *world*; we called it spacetime. Finally, a "curve in a world", or a continuous chain of events, that may correspond, for example, to a moving particle, Minkowski called a *worldline*. These elements are shown in Fig. 5.2.

The worldlines A and B in Fig. 5.2 are straight lines; they correspond to objects moving with constant velocities. From the slopes of these worldlines follows that $v_A = 0$ and $v_B > 0$. The worldline C is curved, its slope is continuously changing, as would be in the case of an accelerating particle.

Where the worldline C intersects the coordinate line x, a flash of light is emitted and propagates in both directions. The electromagnetic signal covers a distance of 1 s each second (in the c-based units), therefore the worldlines of all electromagnetic signal in vacuum are straight diagonal lines. The invariant nature of the speed of light is reflected in the fact that for *all Minkowski diagrams* the worldlines of light are diagonal.

According to Hermann Minkowski, the study of different worldlines and their relationships is of central importance. In his 1908 work *"Space and Time"*,[1] he wrote: *"The whole world presents itself as resolved into such worldlines, and I want to say in advance, that in my understanding the laws of physics can find their most complete expression as interrelations between these worldlines."* The study of interrelation between spacetime curves is, essentially, the study of spacetime geometry.

5.4.2 Drawing Tips

Already with three objects spacetime diagrams may quickly become cluttered, resulting in more confusion than help. It is recommended to draw them *neatly*, using *different colors* to distinguish the worldlines of different objects and light-pulses. In

---

[1] H. Minkowski, *Space and Time*, in *The Principle of Relativity*, Dover, 1952.

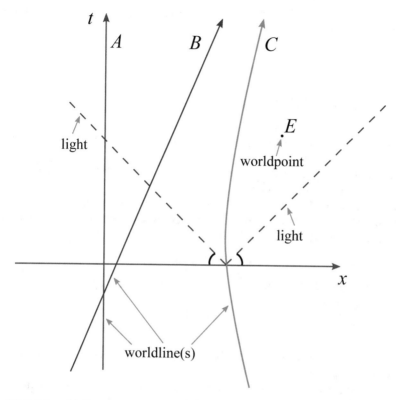

Fig. 5.2 Minkowski diagram and its basic elements: *worldpoint* (event), *worldline* (continuous chain of events), and *world* (spacetime). The diagram also shows the worldlines of light signals using dashed lines. According to the light-postulate, the light lines are *diagonal in all Minkowski diagrams*

the spacetime diagrams in this book, the worldlines of light pulses will be drawn using dashed lines.

? Problem 5.2

Based on the Minkowski diagram in Fig. 5.2, draw another diagram for the same scenario of events, but using the reference frame of the observer B.

? Problem 5.3 *

Spacetime diagrams graphically represent the space of events. Think about what events fill the spacetime diagram of three spaceships in a large intergalactic void? It is not hard to answer this question for the events on the worldlines of the spaceships. You may think that the clock of each spaceship is doing "tick-tock" very rapidly, with each "tick" and each "tock" constituting an event; see Fig. 5.3. But what about events outside the spaceships?

5.5   Relativity of Simultaneity

Using Minkowski diagrams it is possible to demonstrate graphically that two events simultaneous for one observer can not be simultaneous for another one, if the observers are in relative motion. In other words: simultaneity is not an invariant or absolute, it is a relative notion.

Analysis of simultaneity is a good opportunity to practice drawing Minkowski diagrams. The reader should go through the step-by-step exercise below, consulting the solutions given in the Sect. B.19, if necessary.

Step-By-Step Exercise

On two separate sheets of paper, draw two spacetime diagrams describing the same events listed below.

Part A

1. Draw a vertical line for a worldline of an observer M.
2. Draw two more worldlines, for observers A and B, at rest relative to M, and located at equal distances from M in different places.
3. Choose an event E on the worldline of M and draw two worllines of light signal starting at E and propagating in both directions. Draw the light worldlines until they intersect the worldlines of A and B.

Fig. 5.3 Spacetime is "filled with events". What kind of events happen in the apparent void, where there seems to be nothing at all?

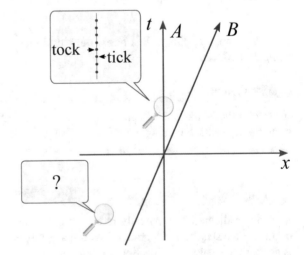

4. Indicate the events, O_A and O_B, where the light from M reaches A and B, respectively. Suppose that the observers A and B set their clocks to zero the moment they receive the signal.
5. Make sure you clearly see that the events O_A and O_B are simultaneous, according to the observer M.
6. Draw the worldline of an observer C moving relative to M, such that C is present at E.

Part B

1. Switch into the reference frame of the observer C and draw its worldline vertically. Mark the event E.
2. Draw the horizontal line, going through E, indicating all events that happen at the same time as E, *according to C*.
3. Draw the wordlines of the observers M, A, and B.
4. Draw the wordline of the light signal emitted by M at E. Indicate the events O_A and O_B.
5. Observe that the events O_A and O_B can not be simultaneous for C. Which event does happen later?
6. * If the last two steps do not show the effect clearly, try to redraw the diagram, assuming that the relative speed of C and M is much closer to the speed of light; in other words, keep the slope of the wordline C closer to the diagonal line.

5.6 Doppler k-factor

The same car-horn sounds differently for a stationary and a moving car. When a car is approaching you, the pitch is higher, compared to the pitch of the stationary horn. For the receding car the pitch is lower. The faster the car is moving, the more pronounced this effect is. This is an example of a *Doppler effect* for sound waves.

In 1842 Christian Doppler suggested that the measured wavelength of light from stars may depend on the their motion relative to the Earth. His conclusion relied on the analogy between water surface waves and light. Three years later, Christoph Buys Ballot experimentally demonstrated how the pitch of the sound from a musical instrument changes when the instrument is moving relative to an observer. In 1848, Hippolyte Fizeau came to the same conclusions as Doppler and proposed that the change in measured wavelength could be visible as the shift of narrow lines in spectra; refer to Fig. 5.9b. Twenty years later, the first experimental confirmation of Doppler shift for light was made by William Huggins. Even more accurate demonstration of optical Doppler effect was made in 1872—three decades after the theoretical

(continued)

prediction of the Doppler effect!—by Hermann Vogel, who measured the shifts of spectral components of light from the Sun due its rotation.

The relative motion of two observers affects *all repeating signals*. Consider, for example, the situation shown in Fig. 5.4a: The observer A is sending two light pulses towards the observer B, while the latter is moving away from A with the speed v. The part Fig. 5.4b shows the Minkowski diagram for this scenario.

If $T_E^{(A)}$ is the time between two emitted pulses, as measured by the observer A, and $T_R^{(A)}$ is the time between received pulses, also as measured by the observer A, then it can be shown that

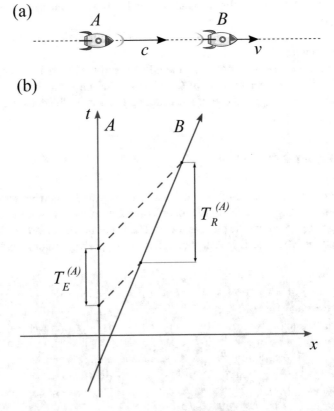

Fig. 5.4 (**a**) Two observers are moving relative to each other. The observer A emits a pair of signals towards B. (**b**) Minkowski diagram of the observer A sending signals to the observer B. The signals will be received by B separated by a longer time interval, compared to the emitted signals

Fig. 5.5 Minkowski diagram of signals propagation from the observer A to the observer B. The observer A sends the signals at t_1 and t_2, and calculates the times T_1 and T_2 when these signals will reach the observer B. Only the times measured by the clock of the observer A are used

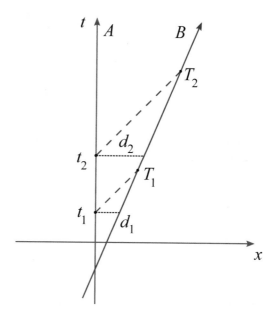

$$T_R^{(A)} > T_E^{(A)}.$$

First, comparing the vertical distance between two emission events to the vertical distance between two reception events, it is clear that $T_R^{(A)} > T_E^{(A)}$; see Fig. 5.4b. Second, we can find the exact relationship between $T_R^{(A)}$ and $T_E^{(A)}$ with the help of the Minkowski diagram and some algebra.

Suppose that at the moment when the observer A emits the first signal, the distance between the observers is d_1, as shown in Fig. 5.5. If the time of emission is t_1, the time of reception is T_1, then the total travel time for the signal is

$$\Delta t_1 = T_1 - t_1.$$

During this time the light signal has to cover the distance d_1 and then some, because the observer B will have moved by $v\Delta t_1$:

$$c\Delta t_1 = \Delta t_1 = d_1 + v\Delta t_1.$$

From the last equation we find Δt_1, and conclude that

$$T_1 = t_1 + \Delta t_1 = t_1 + \frac{d_1}{1 - v}.$$

Similar arguments apply for the second signal. Given the distance d_2 between the observers at the moment of emission of the second light signal, we find

$$T_2 = t_2 + \frac{d_2}{1 - v}.$$

The time between the received signals is

$$T_R^{(A)} = T_2 - T_1 = t_2 - t_1 + \frac{d_2 - d_1}{1 - v} = T_E^{(A)} + \frac{d_2 - d_1}{1 - v}.$$

Clearly, $T_R^{(A)} > T_E^{(A)}$. Between the emission of the first and the second signal the observer B will move by

$$d_2 - d_1 = v(t_2 - t_1) = vT_E^{(A)},$$

and therefore

$$T_R^{(A)} = T_E^{(A)} + \frac{vT_E^{(A)}}{1 - v} = \frac{T_E^{(A)}}{1 - v}.$$

Denoting

$$k^{(A)} = \frac{1}{1 - v},$$

we arrive at the following relationship:

$$T_R^{(A)} = k^{(A)} T_E^{(A)}, \quad k^{(A)} \geq 1.$$

The factor $k^{(A)}$ quantifies the "stretching" effect between the received signals, as compared to the emitted signals.

Important Note
In the expressions above, the superscript (A) indicates that all time measurements refer to the clock of the observer A. In Galileo-Newtonian mechanics such distinction is unnecessary, because the time interval between any two events is an absolute (invariant) quantity. We can not make such an assumption in the special theory of relativity! Recall that two simultaneous events for an observer A, with zero time interval between them, are *not* simultaneous for other inertial observers moving relative to A.

   Thus, in the analysis given above, we avoided comparing the readings of the A's clock to the readings of the B's clock; we relied only on the clock of the observer A. It turns out, if we carry the same analysis in the reference frame of the observer B, we will get a *different* result for the Doppler k-factor!

? Problem 5.4

(a) Draw Minkowski diagram for the same scenario of events from the reference frame of the observer B.
(b) Analyze the problem in the reference frame of the observer B. Show that the k-factor in this case equals $k^{(B)} = 1 + v$.

Which observer is correct?

Discussion

Both observers will agree that the time between the received signals will be greater than the time between the emitted signals.

The observer A finds the "stretch" factor

$$k^{(A)} = \frac{T_R^{(A)}}{T_E^{(A)}} = \frac{1}{1 - v},$$

whereas the observer B obtains a different expression:

$$k^{(B)} = \frac{T_R^{(B)}}{T_E^{(B)}} = 1 + v.$$

Remember that the superscripts (A) and (B) indicate the reference frame to which the result applies.

The expressions for $k^{(A)}$ and $k^{(B)}$ are clearly different. Using the geometric series formula[2]

$$\frac{1}{1 - q} = 1 + q + q^2 + q^3 + \dots$$

we can write

$$k^{(A)} = 1 + v + v^2 + v^3 + \dots = k^{(B)} + v^2 + v^3 + \dots$$

Clearly, $k^{(A)}$ is greater than $k^{(B)}$, but their difference is quite small if the relative speed of the observers is small, i.e. $v \ll 1$. In Galileo-Newtonian mechanics, which is valid for $v \ll c$, the "stretching" between two light signals is almost the same for A and B. Indeed, even for the moderate relative velocity $v = 0.1$, the Doppler k-factors have close values

---

[2] Refer to Appendix A.1.

$$k^{(A)} = 1.11 \text{ and } k^{(B)} = 1.10.$$

It may seem contradictory that two equivalent observers measuring the time intervals between the same pairs of events (two emissions and two detections) get different values. Who is right?

Both are right, as long as they refer to the measurements performed solely in their reference frames. Both A and B use the same speed of the signal, the same relative speed, and the same logic to infer time of distant events. The expression for $k^{(A)}$ involves the times *as measured by the observer A*, while the expression $k^{(B)}$ involves times *as measured by the observer B*. If we assume, as we do in Newtonian mechanics, that the observers A and B measure the same time between two consecutive emissions ($T_E^{(A)} = T_E^{(B)}$) and the same time between two consecutive detection events ($T_R^{(A)} = T_R^{(B)}$), then we would face a contradiction.

The time $T_E^{(A)}$ has a clear physical meaning: the number of ticks the clock carried by the observer A makes between two consecutively emitted pulses. Similarly, the time $T_R^{(B)}$ corresponds to the number of ticks the clock carried by the observer B makes between two consecutively received pulses. Due to the relative motion

$$T_R^{(B)} > T_E^{(A)}.$$

To emphasize, here we compare the readings of two *different clocks*, located in *different places*, and moving relative to each other.

We can look for the value of the following quantity, called *Doppler k-factor*:

$$k = k^{(AB)} = \frac{T_R^{(B)}}{T_E^{(A)}}.$$

This is the ratio of the *number of ticks* that happen on B's clock to the *number of ticks* that happen on A' clock. The number of ticks must be the same for *any* reference frame, therefore the value for k, as defined above, must be the same for both observers A and B (and any other inertial observer!) We will find the expression for this *relativistic* Doppler k-factor in the next step. So far, we can make the following conclusion:

▷ Optical Doppler Effect

When two observers are receding from each other, the time between the received pair of electromagnetic signals T_R will be longer than the time between the emission of those signals T_E. The ratio

$$k = \frac{T_R}{T_E}, \quad k \geq 1$$

depends on the relative speed v of the observers: $k = k(v)$.

5.7 Relativistic Doppler Effect

To find the relativistic expression for the Doppler k-factor in terms of the relative velocity v, we will use the radar method for measuring the distance between two observers (see Sect. 3.4.3). A useful relationship from the radar method connects the time of emission t_E, the time of detection t_D, and the spacetime coordinates (t, x) of the reflection event:[3]

$$t_E = t - x/c = t - x, \quad t_D = t + x/c = t + x.$$

The scenario we will analyze is illustrated in Figs. 5.6 and 5.7. The former shows a series of relative positions of the spaceships at different moments of time, corresponding to the events 1 through 4; the latter shows Minkowski diagram for the reference frame of the spaceship A. It has the following key events:

1. The spaceships A and B meet and set their clocks to zero by exchanging pulses.
2. Some time later, the spaceship A emits the second pulse in order to locate B.
3. The pulse from A gets reflected from B; at the same time the spaceship B sends towards A a pulse of its own (blue dashed-dotted line).
4. The pulse reflected from B is received by A, simultaneously with the other pulse sent by B.

If the observer A measures the position of the event 3 (reflection of the light from the spaceship B) as x and the time as t, then the event 2 (emission) must happen at the time $t - x$; the event 4 (reception) must happen at the time $t + x$ on the A's clock. The spaceship B is moving with the speed v relative to A, and both set their clocks to zero when the meet. Therefore, according to A, the distance between the observers at the time t must be $x = vt$.

The spaceship A sends two pulses to B: the first at the event 1, and the second at the event 2. The time interval between them is $t - x$, according to A's clock. The spaceships A and B are receding from each other, therefore the observer B will receive these pulses separated by the time interval $t_B = k(t - x)$, *according to its clock*. Similarly, the observer B sends towards A two pulses: the first at the event 1, and the second at the event 3. The time interval between them is $t_B = k(t - x)$, according to B's clock. Again, since the spaceships A and B are receding from each other, A will receive the pulses from B separated by the time interval $kt_B = k^2(t - x)$, *according to its clock*. On the other hand, the same time equals $t + x$. We thus have the equality

$$k^2(t - x) = (t + x),$$

from which follows

---

[3] It is assumed in this case that $x > 0$.

Fig. 5.6 Space-only presentation of the scenario of events shown in Fig. 5.7: Observer A uses the radar method to locate a moving object B

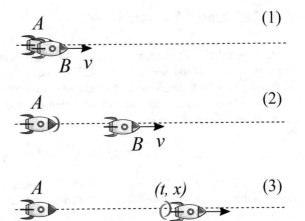

$$k^2 = \frac{t+x}{t-x} = \frac{t+vt}{t-vt} = \frac{1+v}{1-v};$$

where we used the fact that $x = vt$. Finally, we obtain

$$k = \sqrt{\frac{1+v}{1-v}}.\tag{5.1}$$

The behavior of the function $k(v)$ is shown in Fig. 5.8.

? Problem 5.5

Analyze this problem from the reference frame of the observer B.

Hint: You may add one more signal exchange so that B also uses radar method to locate A.

? Problem 5.6

Express the relative speed v in terms of k.

Fig. 5.7 Using the radar method to determine the distance between two receding observers allows one to find the relation between the Doppler k-factor and the relative speed v

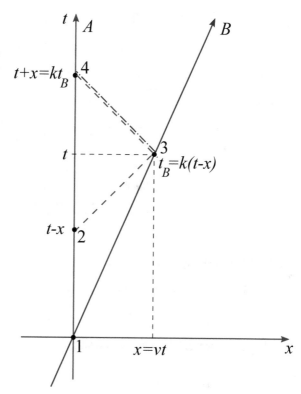

? Problem 5.7

Find the approximate value of the expression

$$z = k - 1$$

for $v \ll 1$.

? Problem 5.8

When an observer A sends a pair of signals to an observer B, the time between the received signals is k times longer than the time between the emitted signals, if the observers A and B are receding from each other. Imagine this situation "played backwards in time", and answer these questions: (1) Which observer is now the emitter? (2) Which observer is the detector? (3) Is the time interval between the received signals longer, shorter, or the same? If it is shorter, how much shorter?

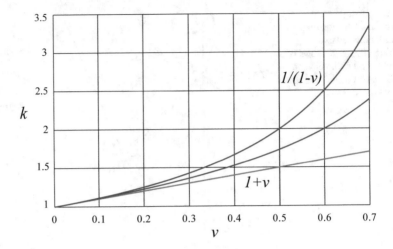

Fig. 5.8 The behavior of the relativistic Doppler k-factor as the function of relative speed v (solid red line). For comparison, the values of the k-factors derived previously relative to the reference frames of the emitter ($k^{(A)}$, blue dotted line) and the receiver ($k^{(B)}$, straight green dashed-dotted line)

5.7.1 Doppler Shift and Astrophysics*

The effect discussed above is of tremendous importance for astrophysics. Electromagnetic radiation from stars, galaxies, gas clouds, quasars, and other sources contains information about their material composition, internal processes, and motion relative to the observers on Earth. Astronomical spectroscopy studies *spectra* of electromagnetic radiation from astronomical objects. A spectrum is a way to represent radiation in terms of various "colors" (wavelengths or frequencies), showing how much energy a particular wavelength or frequency contributes into the total energy of the radiation.

Figure 5.9 shows spectra of the Sun and hydrogen—the most common element in the universe. When a beam of light from the Sun goes through a prism, it *disperses* into various colors, including the visible part of the spectrum from blue and violet (shorter wavelengths) to red (longer wavelengths). The energy of radiation from the Sun is smoothly distributed between different wavelengths, with the maximum near the yellow color (580 nm) as shown in Fig. 5.9a. Such a spectrum is characteristic to a very hot body.

In contrast, electromagnetic radiation from a cloud of hot hydrogen atoms contains a discrete set of components. Several of them, belonging to the visible part of the spectrum, are shown in Fig. 5.9b. Every chemical element has a distinct set of components, each located at a specific place in the spectrum. A careful analysis of the spectrum can reveal the composition of the source of radiation.

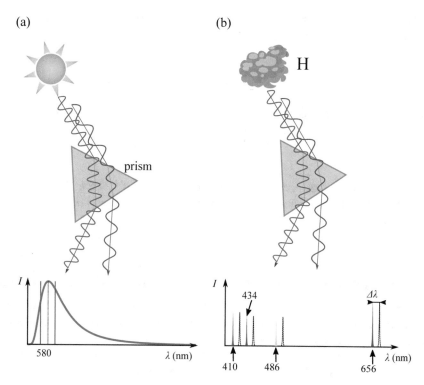

Fig. 5.9 Different types of spectra: (**a**) Continuous spectrum of a hot body (sun). Two vertical lines indicate the bounds of the visible part of the spectrum: from violet to red. (**b**) Discrete spectrum of an atomic hydrogen. Three lines from visible part of the spectrum (434 nm, 486 nm, and 656 nm) and their "red-shifted" copies are shown as unshaded peaks. The shift of each line is characterized by the magnitude $\Delta\lambda$. Hydrogen has many more lines outside of the visible part of the spectrum

When the source of radiation is moving relative to the observer on earth, the positions of the spectral components will be shifted relative to their position as measured in the laboratory; see unshaded "copies" of the peaks in Fig. 5.9b. For a source moving away, the shift will be towards longer wavelength (*red shift*), whereas for a source approaching the laboratory, the shift will be towards the shorter wavelengths (*blue shift*). The magnitude of the shift $\Delta\lambda$ can be used to determine how fast the distance between the radiation source and the Earth is changing.

? Problem 5.9

The wavelength of an electromagnetic radiation is proportional to the period T of a single oscillation of electric charges in the source:

$$\lambda = cT.$$

This allows expressing the Doppler factor k in terms of the ratio of the wavelengths of the received and the emitted signals:

$$k = \frac{T_R}{T_E} = \frac{\lambda_R}{\lambda_E}.$$

A commonly used quantity in astrophysics is the *relative shift of a wavelength*

$$z = \frac{\Delta\lambda}{\lambda} = \frac{\lambda_R - \lambda_E}{\lambda_E}.$$

Find the relation between z and the relativistic Doppler factor k. Then find the relative velocity for the following values of z: (1) $z = 0.01$; (2) $z = 0.1$; (3) $z = 1.0$; (4) $z = 10.0$.

? Problem 5.10

Electromagnetic radiation detected from different galaxy clusters shows the presence of ionized calcium; see Fig. 5.10. Two close spectral lines (K and H) are detected as a single peak near 395 nm. For more distant clusters, such as Gemini cluster, the signal is weaker, and the position of the peak is red-shifted to 425 nm. Find the speed of this cluster relative to an observer on Earth.

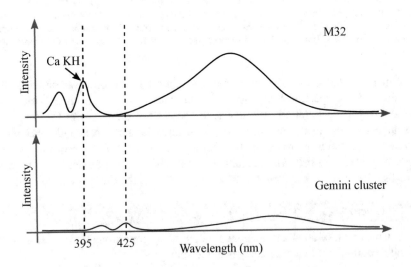

Fig. 5.10 Spectra of electromagnetic signal from a nearby (M32) and more distant (Gemini) galaxy clusters. Both clusters show the features attributed to ionized calcium (K and H lines), but in different positions on the wavelength axis. We can use the Doppler effect to find the speed of each cluster relative to an observer on Earth

5.8   Muon Experiment

Consider the following two imaginary (but not impossible) scenarios. First, imagine you have a candle that lasts for exactly 1 h. If you place this candle into a car that travels with the speed 100 km/h, it will be 100 km away by the time the candle burns out. This reasoning is wrong; in fact, the candle will travel 100 km and, in addition, about 6 millimeters. If the candle is placed into a spaceship traveling with the speed $v = 0.87$, then the distance covered will be *twice* the value expected from the Galileo-Newtonian mechanics.

Second, imagine you get on a fast magnetically levitating train. When the train leaves the station A, your clock shows the same time as the station's clock—exactly zero hours. When you arrive at another station B, in the *same* time zone, you see that the station's clock shows exactly 10 h. If the train traveled with the speed 400 km/h, your clock will be 32 milliseconds short of 10 h. Using the spaceship with $v = 0.87$, the clock of the traveler will indicate *half* of the time elapsed on the stations' clocks.

The effects similar to the described above are observed every day in large numbers around the world. Many elementary particles, created in collision experiments, travel with speeds comparable to the speed of light. In addition, many elementary particles are *unstable*—they decay after some time, turning into other particles, analogously to the neutron decay discussed in Sect. 3.3.5.

Muon is an elementary particle similar to electron, but 207 times heavier and, in contrast to electron, is unstable. Muons can be produced in a laboratory, but they also come from a natural collider—cosmic rays entering the Earth's atmosphere, as illustrated in Fig. 5.11a.

On average, a muon *at rest* will decay after $\bar{\tau} = 2 \times 10^{-6}$ (s) (two millionth of a second or two microseconds). From the point of view of Newtonian kinematics, the maximum possible distance a muon can travel is

$$\bar{d} = \bar{c}\bar{\tau} = 3 \times 10^8 \, (m/s) \times 2 \times 10^{-6} \, (s) = 600 \, (m).$$

This is much smaller than the distance to the upper atmosphere of $D > 10000$ meters. However, a steady flux of high speed muons is observed at the sea level and below.

5.8.1   Frisch and Smith Experiment

In 1963 David H. Frisch and James H. Smith performed an experiment,[4] where they measured the number of muons of certain energy detected within 1 h. The experiment was filmed and available as a short movie *"Time Dilation—An*

---

[4] David H. Frisch and James H. Smith, *"Measurement of the Relativistic Time Dilation Using μ-Mesons,"* American Journal of Physics 31 (1963), 342–355.

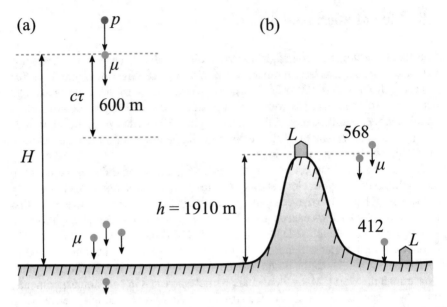

Fig. 5.11 (a) Muons are produced in collisions between high energy cosmic protons and molecules in the atmosphere. They reach the surface of the earth and even penetrate into deep mines. (b) In the muon experiment of David Frisch and James Smith, the number of muons detected during a fixed time interval was measured at two different altitudes. The measurements at the foot of the Mt. Washington revealed much higher detected number of muons than expected from Newtonian physics

Experiment With μ-Mesons", by Educational Services Incorporated.[5] Figure 5.11b illustrates the idea of the experiment.

Frisch and Smith first measured the number of muons at the top of the mountain Washington in New Hampshire, United States. On average, 568 muons per hour were stopped by their detector, all having the velocities between 0.9950 and 0.9954.

Additionally, Frisch and Smith measured the muon's average time to decay after they have been stopped. It was equal to 2.2×10^{-6} s. If these muons had not been stopped, they would have traveled less than

$$0.9954 \times 3 \times 10^8 \, (m/s) \times 2 \times 10^{-6} \, s = 657 \, (m),$$

according to Newtonian kinematics. Of course, there is a chance that some muons could survive a much longer travel time and distance. However, this chance is low and a negligible number of such muons should reach the sea level.

In the second part of the experiment, the laboratory was moved down to the sea level (about 1910 meters down), the experiment was repeated and the average

---

[5] Available on Youtube.

Fig. 5.12 A muon travels
from the top of the mountain
to the sea level with $v \approx 1$.
When it reaches the sea level,
its "clock" would read much
less time than the clock of the
laboratory

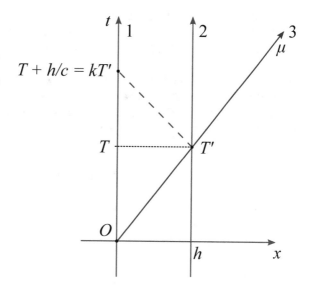

number of muons from the same group of velocities was measured. The average
number of muons became 412 muons per hour—not significantly lower than
before. This result goes against the predictions of pre-relativistic Newtonian physics.
However, using Minkowski diagram and the expression for the relativistic k-factor
we can show that this is exactly what should happen.

5.8.2   Analysis of the Experiment

Consider Minkowski diagram in Fig. 5.12. It corresponds to the reference frame
of the mountain and shows the following worldlines: (1) The worldline of the
laboratory on the top of the mountain, where a muon starts its journey at time $t = 0$;
(2) the worldline of the laboratory at the sea level, situated h meters away from the
top; (3) the worldline of a muon moving from the top to the bottom with the speed
$v = 0.9950c$. The axis x is chosen to point in the direction from the top of the
mountain to the sea level.

If the muon leaves the top of the mountain at time $t = 0$, it will arrive at the sea
level at time $T = h/v$, according to the clocks of both laboratories at the top and
the bottom. Let's imagine that the third clock is moving together with the muon, and
denote the time on this clock when the muon reaches the sea level as T'. Further,
imagine that the laboratory at the sea level sends a light signal at the moment when
the muon reaches it. The signal will arrive to the top at the moment

$$T + \frac{h}{c} = T + h,$$

according to the clocks of both laboratories.

The diagram Fig. 5.12 looks exactly like the diagram of a signal exchange between two receding observers. The emitter, worldline 3, sends two signals to the detector, worldline 1. The first light signal is emitted at $t = 0$ and is received instantly. The second light signal is sent at the moment T', according to the clock of the emitter. The time between two emitted signals $T_E = T'$, according to the clock of the emitter, while the time between the received signals is $T_R = T + h$, according to the clock of the receiver. As we showed, they are related via relativistic Doppler k-factor

$$T + h = kT'.$$

Given the speed of the muon v, we can find

$$h = vT$$

and

$$T' = T\frac{1+v}{k} = T\sqrt{1 - v^2}. \tag{5.2}$$

From this relation follows that $T' \leq T$ for all velocities, and $T' \ll T$ for fast muons. Using $v = 0.9950$, we get

$$T' = 0.01T.$$

Thus, when the clocks in both laboratories indicate

$$T = \frac{\bar{h}}{\bar{v}} = \frac{1910\,(m)}{0.9950 \times 3 \times 10^8\,(m/s)} = 6.4 \times 10^{-6}\,(s),$$

the clock moving along with the muon would measure

$$T' = 0.064 \times 10^{-6}\,(s).$$

For the observer in the laboratory, the time T of muon's travel from the top to the bottom of the mountain is greater than the muon's lifetime. Galileo-Newtonian mechanics predicts that almost no muons should survive this long, and hardly any muon should reach the sea level. The observer at rest relative to the muon (co-moving clocks) will measure time T' much smaller than the muon's lifetime. According to this observer, it is highly unlikely that any muon decays before reaching the sea level.

Different time measurements for different observers may appear in conflict. However, the observers in the laboratories should not expect fast muons to decay before they reach the sea level. It is important to remember that the decay time of 2.2 microseconds is measured for muons *at rest* relative to the clocks. It turns out, the assumption that moving clocks remain in sync with stationary clocks *is wrong*, it is not supported by experiments. In our daily lives, the effects of motion on time measurements are tiny and unnoticeable by our senses. Our brains make unjustified conclusion that identical clocks always remain in sync, regardless of their relative motion. The first conclusion we can draw from the muon experiment can be expressed as follows:

▷ **Property of Moving Clocks**

Clocks in relative motion do not remain synchronized.

? Problem 5.11

Alpha Centauri—the star system closest to the Sun—is $d = 4.37$ years away. How fast should a delivery spaceship travel in order to deliver flowers in 2 days time, according to the spaceship's clock?

? Problem 5.12

Analyze the scenario from Fig. 5.12 from the reference frame of the muon.

5.9   Composition of Velocities

When discussing the relative nature of velocities in Galileo-Newtonian mechanics, we showed that velocities are simply added. Specifically, if we have three observers, like shown in Fig. 5.13a, then to find the velocity of the observer C relative to A, we simply add the velocity of C relative to B and the velocity of B relative to A:

$$w = u + v.$$

The formula works well, provided $v \ll c$ and $u \ll c$.

To find the relativistic formula connecting the velocities v, u, and w, we can use Minkowski diagram shown in Fig. 5.13b. The diagram corresponds to the reference frame of the spaceship A.

Any two pulses, emitted by the spaceship A separated by T seconds, according to the A's clock, will reach the spaceship B separated by $T' = k_1 T$ seconds, according to the B's clock. Here

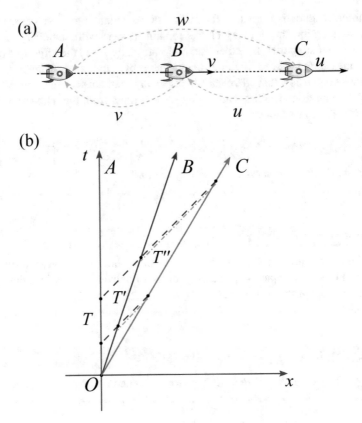

Fig. 5.13 (a) The observer C is moving relative to B, while B is moving relative to A. What is the speed of C relative to A? (b) Using the method of Doppler k-factor, we can determine the rule for the composition of velocities for several observers moving relative to each other

$$k_1 = k(v) = \sqrt{\frac{1+v}{1-v}}$$

is the relativistic Doppler k-factor for the observers A and B.

Imagine that when a pulse from A reaches the observer B, the latter instantly emits a pulse of its own towards the observer C. Such signals are shown in Fig. 5.13b as blue dashed-dotted lines running parallel to the worldlines of the pulses from A. The signals from B travel together with the pulses from the spaceship A. Since the spaceships B and C are receding from each other, the pulses emitted by B will arrive to C separated by $T'' = k_2 T'$ seconds, according to the C's clock. Here

$$k_2 = k(u) = \sqrt{\frac{1+u}{1-u}}$$

is the relativistic Doppler k-factor for the observers B and C.

Therefore, the pulses from the spaceship A are received by the spaceship C separated by

$$T'' = k_2 T' = k_2 k_1 T = k_3 T \text{ seconds.}$$

Here

$$k_3 = k(w) = \sqrt{\frac{1+w}{1-w}}$$

is the relativistic Doppler k-factor for the observers A and C. We derived an important and useful result:

- k-factors for several relative motions are simply multiplied. In the case of three observers we find

$$k_3 = k_2 k_1.$$

To find the relationship between the velocities, note that

$$w = \frac{k_3^2 - 1}{k_3^2 + 1},$$

and

$$k_3^2 = k_1^2 k_2^2 = \frac{(1+u)(1+v)}{(1-u)(1-v)}.$$

Opening the parentheses in the last expression results in

$$k_3^2 = \frac{1 + u + v + uv}{1 - u - v + uv}.$$

Next we find

$$k_3^2 - 1 = \frac{2(u+v)}{1 - u - v + uv},$$

and

$$k_3^2 + 1 = \frac{2(1+uv)}{1 - u - v + uv}.$$

Finally, we arrive at the velocity composition rule:

$$w = \frac{u+v}{1+uv}. \tag{5.3}$$

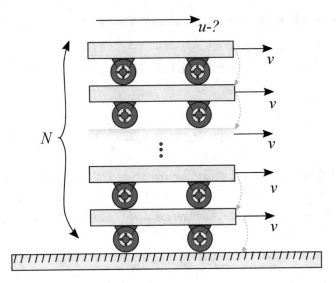

Fig. 5.14 What is the speed of the top car relative to the ground, given that the speed of each cart relative to the one below is v?

When both u and v are much smaller than the speed of light, the product uv is even smaller and can be neglected. In this case the composition rule reduces to simple addition of velocities.

? Problem 5.13

N carts are stacked on top of each other, as shown in Fig. 5.14. The speed of the first cart (at the bottom) relative to the ground is v, the speed of the second cart *relative to the first* is also v, the speed of the third cart *relative to the second* is also v, and so on. Find the speed of the topmost cart *relative to the ground*.

5.10 Blue Shift and Reciprocity

Earlier we showed that when two observers are receding from each other, the time between a pair of received signals will be longer than the time between their emission. This is illustrated in Fig. 5.15a, where the observer B is moving away from the observer A. We derived the relationship

$$T' = kT, \quad k = \sqrt{\frac{1+v}{1-v}}, \ k \ge 1.$$

Fig. 5.15 (**a**) Signals from the observer A are received by two other observers, B and C. The observer B is receding, while the observer C is approaching. (**b**) The relationship between the "stretching" and "compression" of time interval between a pair of signals, as measured by the receding and approaching observers

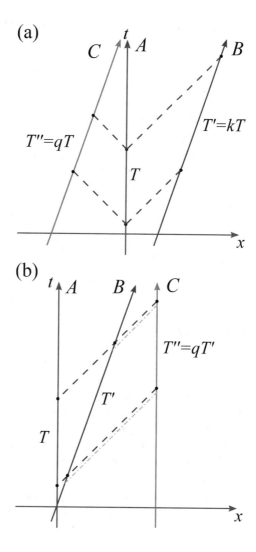

The observer C, shown in Fig. 5.15a is moving *towards* A with the same velocity as B. The slope of the worldline C is the same as that of B. We can show that the time between a pair of signals received by C is related to the time between their emissions as

$$T'' = qT, \quad q \le 1,$$

where q is the Doppler factor for approaching observers. The value of this factor can be found following arguments similar to the ones used previously to find $k(v)$. The answer can also be anticipated if one considers the relative motion of the observers A and B, from Fig. 5.15a, "backwards in time": A and B will be approaching each other, the observer B becomes the emitter, and the observer A becomes the detector. There is, however, yet another simple way to get the result.

Consider the scenario shown in Minkowski diagram in Fig. 5.15b. The diagram corresponds to the reference frame of the observer A. The observer C is at rest relative to A, some distance away. The observer B is moving from A to C. The observer A sends two pulses separated by T seconds, according to A's clock. They reach the observer B separated by time interval $T' = kT$, according to B's clock.

Imagine that every time a pulse from A reaches B, the latter emits a pulse of its own towards C. Such signals are shown in Fig. 5.15b using blue dashed-dotted lines, running parallel to the worldlines of the pulses from A. The signals from A and B will reach the observer C at the same time; they will be separated by the time interval

$$T'' = qT' = T,$$

where we used the relationship between the emitted and recevied signal for the approaching observers. Using the fact that $T' = kT$, we obtain

$$qkT = T,$$

consequently,

$$q = \frac{1}{k}.$$

The time interval between a pair of signals will become "shorter" when the emitter and the receiver are approaching each other. The magnitude of this effect is quantified by the Doppler factor $q(v)$, which is the inverse of the Doppler factor $k(v)$ used for two observers receding from each other with the same speed v.

The results about the relativistic Doppler k-factors can be summarized as follows:

- When two observers are receding from each other with the relative speed v, the time T_R between the received signals will be longer than the time T_E between the emitted signals:
$$T_R = kT_E, \quad k \geq 1.$$

- When two observers are approaching each other with the relative speed v, the time T_R between the received signals will be shorter than the time T_E between the emitted signals:
$$T_R = qT_E, \quad q = \frac{1}{k}, q \leq 1.$$

- The value of the k-factor depends on the relative speed of two observers
$$k(v) = \sqrt{\frac{1+v}{1-v}}.$$

5.11   Path Dependence of Time

The muon experiment, discussed in the Sect. 5.8, showed that moving clocks and stationary clocks measure time differently. To understand why this happens we must learn an important property of all clocks:

▷ **Function of Clocks**

A clock measures the length of its worldline.

Clocks are spacetime odometers, so to speak. The experiments discussed next will demonstrate this fact.

The first experiment is illustrated in Fig. 5.16. Observers A and B are initially at the same place; they synchronize their clocks and the observer B starts moving with the speed v. When the clock of B indicates time T, the observer turns around and travels back to A with the same speed v.

The observer B is first receding from A; then, after the turn, the observer B is approaching A. Suppose B sends three signals to A: the first signal is emitted at the moment of departure, the second signal is sent at the moment of the turn, when the clock of B measures time T, and the third—at the moment of the second meeting of A and B; see Fig. 5.16b.

The time between the first and the second signal, as measured by the observer A, equals $kT > T$ (receding emitter-detector); the time between the second and the third signal is $T/k < T$ (approaching emitter-detector). The total time on A's clock is then

$$t_A = kT + \frac{T}{k} = T\left(k + \frac{1}{k}\right).$$

Using the expression for k in terms of relative velocity v, the expression in the parentheses can be written as

$$k + \frac{1}{k} = \frac{2}{\sqrt{1 - v^2}}.$$

The time between the first and the second meeting of the observers A and B, according to A, equals

$$t_A = \frac{2T}{\sqrt{1 - v^2}}.$$

The observer B, traveling the same distance with the same speed, measures the round-trip time as

(a)

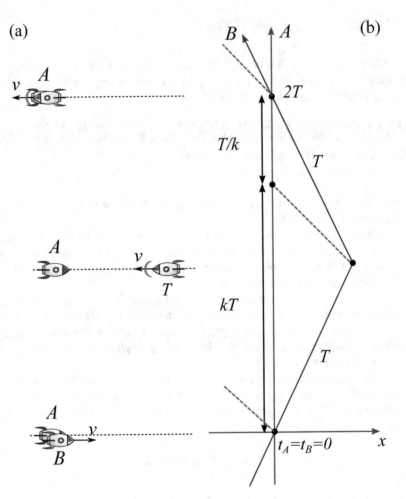

Fig. 5.16 The observer B is making a round-trip, traveling time T in each direction relative to A. (a) Spatial configurations for the key events (from bottom to top): synchronization, turn, the second meeting. (b) Minkowski diagram of the process. Three signals, sent from B to A, are shown as dashed line segments

$$t_B = 2T,$$

which is less that t_A.

Can this be explained by the fact that the observer B is *not inertial*? After all, it has to make a turn in the middle of the journey, and changing the velocity requires an acceleration. The following analysis of a similar scenario, not involving accelerating observers, demonstrates that the phenomenon has nothing to do with acceleration.

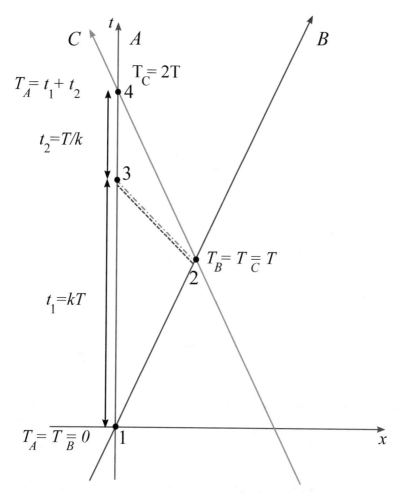

Fig. 5.17 Path dependence of time: The time between any two events measured by a clock depends on the path taken between these events in spacetime. Here, the observer B is moving away from A with the speed v, while the observer C is approaching A with the same speed v. When B and C meet, the latter sets its clock to read the same time T as the clock of B. When C reaches A, the reading on their clocks will be different

Minkowski diagram in Fig. 5.17 shows a scenario of events from the reference frame of an observer A. In this scenario two observers, B and C, are moving relative to the observer A with *equal speeds* in the opposite directions.

The key events are the following:

1. The spaceship B is moving past A; at the moment they meet, both set their clocks to zero (event 1). They also exchange pulses, which happens almost instantly since they are close.

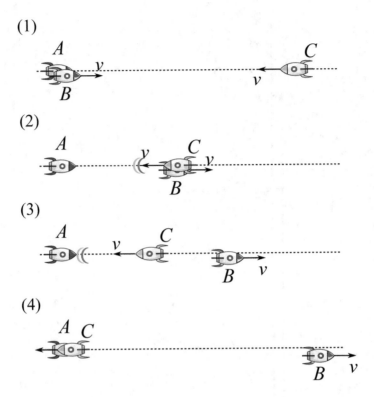

Fig. 5.18 Spatial configurations of the observers involved in the thought experiment of "clock round-trip": (1) The observers A and B set their clock to zero. (2) The observers B and C meet and C sets its clock to the same time T that B's clock shows. (3) The observer A "sees" the meeting of B and C. (4) The observer C passes by A and both compare their clocks to find different time

2. The spaceships B and C meet; C sets its clock to read the same time as the clock of B (event 2). Both spaceships B and C send pulses towards A.
3. The second pulse from the spaceship B and the first pulse from the spaceship C are received at the same time by A (event 3).
4. Finally, the spaceships A and C meet; they compare the readings of their clocks (event 4). They also exchange pulses, which happens almost instantly since they are close.

An alternative presentation of these events is shown in Fig. 5.18.

It is important that the spaceships B and C are moving with the same speed v, although in different direction. Since all observers in the special theory of relativity are inertial, we can not let the observer B decelerate, turn around, accelerate and come back to A. The role of the observer C is to take the "time" of B's clock and "bring it back" to A for comparison.

When the spaceships B and C meet, and both send the pulses to A (event 2), the clock of B reads time T; this is the time the spaceship C uses to set its clock. The observer C is traveling with the same speed as B, therefore the travel to A from the meeting with Bs will require the same time T, according to the clock of C. The latter then must read time $T_C = 2T$ at the event 4, when the spaceships A and C meet.

The time T_A, elapsed on the clock of A between the meeting with the spaceships B and, later, with C, can be split into two parts, as shown in Fig. 5.17. The first part is the time t_1 between the meeting with the spaceship B (event 1) and the detection of the second pulse from B (event 3). Since the spaceships A and B are receding, this time equals $t_1 = kT$. The second part is the time t_2 between the arrival of the first pulse from the spaceship C (event 3) and the meeting of the spaceships A and C (event 4). Since A and C are approaching each other, this time equals $t_2 = T/k$. Thus, when the spaceships A and C meet, the clock of A will measure the total time

$$T_A = t_1 + t_2 = (k + 1/k)T.$$

Using the expression for the k-factor in terms of the relative speed v, we can write the time on the clock of the spaceship A as

$$T_A = \frac{2T}{\sqrt{1 - v^2}} = \frac{T_C}{\sqrt{1 - v^2}},$$

which is *more* than the reading on the clock of the spaceship C. For example, if $v = 0.87$ then $T_A = 2T_C$.

The expression

$$\frac{1}{\sqrt{1 - v^2}}$$

appears often in books on special relativity; it is traditionally denoted using Greek letter gamma:

$$\gamma = \frac{1}{\sqrt{1 - v^2}}.$$

The behavior of γ as the function of relative speed v is discussed in the Appendix A.3. There it is also shown that for relative speeds $v \leq 0.1$ the value of γ can be accurately approximated by the quadratic function

$$\gamma \approx 1 + \frac{v^2}{2}.$$

(continued)

The factor γ—sometimes called *Lorentz factor*—has no special status, unlike the constants π or e in mathematics. It is simply a single letter shorthand for a more complicated expression.

? Problem 5.14

Show that the relativistic factor γ can be expressed in terms of the Doppler k-factor as follows:

$$\gamma = \frac{1}{2}\left(k + \frac{1}{k}\right).$$

? Problem 5.15

Find the time coordinate of the event 2. In other words, at what time, according to the A's clock, the observers B and C meet.

For a racing car moving with the speed 100 meters per second (360 kilometers per hour or 224 miles per hour) the total travel time $2T$ must be about 571,776 years in order to result in the time difference

$$\Delta T = T_A - T_C$$

of just 1 second! If the total travel time $2T$ is not astronomically large, the time difference ΔT is vanishingly small. This is the reason for thinking that synchronized clocks remain synchronized, regardless of their relative motion.

We showed that the clock of the spaceship C will measure time greater than the clock of the spaceship A. The inequality of time measured by different clocks in relative motion may be perceived as a paradox, but as Richard Feynman said:[6] *[Otherwise] the paradox is only a conflict between reality and your feeling of what reality "ought to be".*

Our expectations are based on the experience limited to very low speeds ($v \ll 1$) and relatively short times T; they have nothing to say about how nature behaves outside of this range of experience.

---

[6] R. Feynman, R. B. Leighton, M. Sands *Feynman Lectures on Physics* VIII, Section 18–3.

The fact that the observers A and C will have different time on their clocks when they meet is an *absolute fact*—true for any reference frame. The Exercise 5.16 shows this for the observers B and C. Although the observers A, B, and C are all inertial and equivalent from the standpoint of the laws of physics, their "paths" in spacetime are not equivalent, as is clear from Minkowski diagram Fig. 5.17. Since clocks measure the lengths of their worldlines, it must be not surprising that clocks with different worldlines have different readings.

? Problem 5.16

(a) Analyze the problem considered in Fig. 5.18 in the reference frame of the observer B.
(b) Analyze the same problem in the reference frame of the observer C.

5.11.1 Path Dependent Quantities

One of the important lessons of the special theory of relativity is that time measurements are more similar to distance measurements than was previously thought. Two identical cars traveling between the same cities via different routes may have different total distance traveled. This is an elementary fact, familiar to all. The analysis of the Minkowski diagram in Fig. 5.17 demonstrates that *two identical clocks "traveling" between the same events via different spacetime "routes" may have different total times measured*. This similarity between the measurements of time and distance is highlighted in Fig. 5.19.

Some physical quantities, e.g. the potential energy, do not depend on the route taken by the cars. The change of potential energy will be the same if the cars start from the same point and arrive at another point, as illustrated in Fig. 5.20a. Electrostatic potential energy also has this property.

Whether time is more like potential energy or more like the total distance traveled should be decided by experiments. According to the special (and general) theory of relativity, time is a *path dependent quantity*.

Looking at the spacetime diagram in Fig. 5.17, we see that the worldlines connecting the events 1–4 and 1–2–4 trace different paths in spacetime. The "lengths" of these paths—the time measured by the clocks traveling along them— are different. However, the straight path 1–4 corresponds to the *longer* path, in

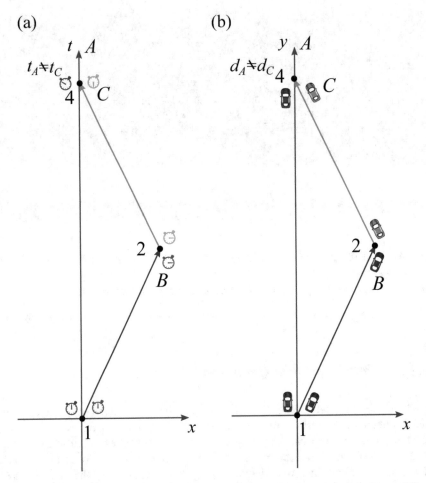

Fig. 5.19 (a) Time between two events measured by a clock depends on the clock's worldline connecting these events. (b) A familiar property of traveling through space—different paths have different lengths

contrast with the properties of lengths in Euclidean geometry. The geometry of spacetime is *non-Euclidean*, as we will soon find.

The path dependent nature of time helps understand deeper the difference between the clocks of the laboratories and the clock moving along with the muon, discussed in the Sect. 5.8. Figure 5.21 shows a simplified version of the Minkowski diagram for the muon experiment. The worldline of a muon, having the "length" T', clearly differs from the worldlines of the clocks in the laboratories, which both have the "length" T. The clocks measure the lengths of their worldlines, therefore the reading of the clocks are different for different worldlines.

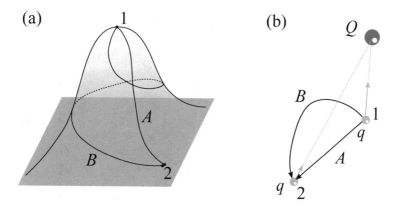

Fig. 5.20 (**a**) Change in mechanical potential energy does not depend on the path (A or B) taken between two points (1 and 2). (**b**) The energy required to move a charge q relative to another charge Q does not depend on the path (A or B) between the initial and final point (1 and 2, respectively)

Fig. 5.21 The clocks in the laboratories and the clock moving along with the muon have different worldlines with different "lengths". Clocks are devices that measure these lengths, hence the difference in the clocks' readings

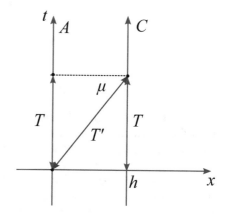

5.12 Looking at Moving Clock

So far the clock of a moving observer was compared to the clocks of stationary observers at the moments when a pair of observers were next to each other. What would we *see* if we were looking at a moving clock, as it is either approaching or receding from us? The problem is much more complicated than it may appear. The answer depends on the shape and color of the clock, and, more importantly, at how we are looking at it. Below we will consider a very simple case.

Consider the situation illustrated in the spacetime diagram in Fig. 5.22. Spaceships A and C are at rest relative to each other, while the spaceship B is moving from A to C. The spaceship B sends a series of pulses, separated by the time interval T, according to its clock, and the pulses propagate in both directions.

Fig. 5.22 Two observers are "looking" at the clock of the third observer, while the latter is either receding (A and B) or approaching (B and C)

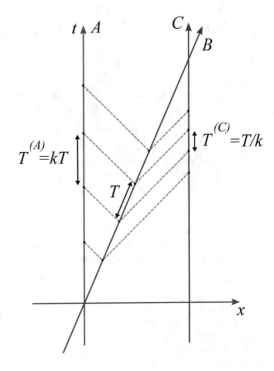

As was shown earlier, the time interval between the pulses received by the spaceship A will be $kT > T$ seconds apart, since A and B are receding. The time interval between the pulses received by the spaceship C will be $T/k < T$ seconds, since B and C are approaching. If the pulses carried images of the face of the clock of the spaceship B, then the spaceship A would "see" that the B's clock was running slow. The spaceship C would "see" the opposite—the clock of the moving spaceship B would seem to run faster that the clock of C. If, for example, the speed of the observer B is $v = 0.6$, then the relativistic Doppler k-factor equals 2; the observer A will see the B's clock advance one second for each 2 s of the A's clock, while the observer C will see that the B's clock advance 1 s every half a second of C's clock.

The apparent "slowing down" or "speeding up" of a moving clock is the familiar Doppler effect for light, related to the red shift and blue shift. Often in astrophysics one needs to determine the speed of a moving source of a signal, given the measured frequency of the signal and *known* frequency of a similar signal from a stationary source. In the experiment considered above, the spaceships A and C know the relative speed and may be interested in the frequency of the emitted train of pulses, or, equivalently, the time T between the emitted pulses, as measured by the clock of B (the emitter).

There are infinitely many possible *measured* frequencies, depending on the relative speed of two observers. However, there is *one* frequency that all observers can infer from their measured frequency—the frequency of a stationary source or *proper frequency*. The proper frequency can be found if the Doppler effect is taken into account. The observer A first needs to find the time between the emitted pulses, as measured by B:

$$T = \frac{T^{(A)}}{k},$$

and then calculate the proper frequency $f_0 = 1/T$. Similarly, the observer C first calculates the time T:

$$T = kT^{(C)},$$

and then the proper frequency f_0. This way both A and C would be able to tell that the signal is coming from the same source, despite the difference of the measured frequency of the received signal.

Common Pitfall

According to the special theory of relativity, identical clocks in different states of relative motion measure time differently. This is an experimentally confirmed fact, pointing to the path dependent nature of time measurements.

In the muon experiment, the clock moving along with the muon measured less time than the clocks in the laboratories. In the round-trip experiment, discussed in the Sect. 5.11, the combined time of the clocks of the observers B and C was less than the time measured by the observer A. It may seem that in both cases the moving clocks were "running slow", leading to less measured time. A question can be asked: *Why and how does motion affect clocks?*

The principle of relativity, both Galilean and special, clearly states that *uniform rectilinear motion does not affect any physical phenomena, including the ones that determine the running of a clock*. For clocks in relative motion, for example the clock co-moving with a muon and the clocks in laboratories, neither is "truly/absolutely moving"—*motion is relative*. If one is "running slow" for some reason, then the other must also be "running slow" in the same way for the same reason; therefore, this is not a suitable explanation for the muon experiment or the round-trip experiment.

The thought that "moving clock is running slow" is common enough that Taylor and Wheeler in their book *"Spacetime Physics"*[7] provide the following comment:

▷ **Taylor and Wheeler on Clocks**

Does something about a clock really change when it moves, resulting in the observed change in tick rate?

---

[7] E. Taylor, A. Wheeler *Spacetime Physics*, Freeman and Company, 2nd ed., p. 78.

Absolutely not! Here is why: Whether a free-float clock is at rest or in motion is controlled by the observer[...]How can your change of motion affect the inner mechanism of a distant clock? It cannot and does not.

Instead of phrases like "moving clocks are running slower," it is better to emphasize the path dependent nature of time measurements.

5.13   Lorentz Transformation

Now we are ready to derive the relationships between the spacetime coordinates of any event, as measured by two different observers in relative motion. The scenario we will analyze is shown as several steps in Fig. 5.23 and its Minkowski diagram in Fig. 5.24. There, two observers, A and B, are using the radar method to locate an object C.

The key events in the Minkowski diagram are the following:

1. The spaceships A and B meet and set their clocks to zero by exchanging light-pulses. This happens almost instantaneously, since the spaceships are next to each other.
2. Some time later, the spaceship A sends a pulse to locate the object C.
3. The pulse from the spaceship A catches up with the spaceship B. At this moment the latter emits a pulse of its own (point-dashed line) to locate the object C.
4. Pulses from A and B are reflected from the object C;
5. The reflected pulses meet the spaceship B on their way towards A.
6. Both reflected pulses are detected by the spaceship A.

The spacetime coordinates of the event 4, measured by the spaceship A, are (t, x); the spacetime coordinates of the same event, measured by the spaceship B, are (t', x'). According to the A's clock, the times of the emission and the reception of the light pulse are $t - x$ and $t + x$, respectively. Similarly, according to the B's clock, the times of the emission and the reception of the light pulse are $t' - x'$ and $t' + x'$.

The spaceship A sends two pulses towards B: the first pulse at the event 1, time $t = 0$; the second pulse at the event 2, time $t - x$. The first pulse is instantly received by the spaceship B at the event 1, time $t' = 0$; the second pulse is received at the event 3 at time $t' - x'$. Since the spaceships A and B are receding from each other, the time interval between the emitted pulses is related to the time interval between the received pulses as

$$(t' - x') = k(t - x). \tag{5.4}$$

When both reflected pulses are passing the spaceship B (event 5) at time $t' + x'$ on its clocks, we may view consider them as if they are emitted by B at that moment.

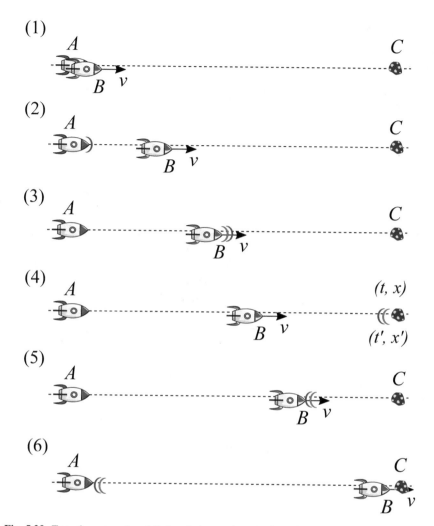

Fig. 5.23 Two observers, A and B, in relative motion use the radar method to locate an object. The reflected signals locate the object at time t and position x for the observer A, and at time t' and position x' for the observer B. Steps 1 through 6 are explained in the text

Therefore, the observer B effectively sends two pulses towards the spaceship A. The first pulse is emitted at the event 1 at $t' = 0$; the second pulse is emitted at the event 5 at $t' + x'$. The first pulse is received instantly by the spaceship A at the event 1 at time $t = 0$, the second pulse is received at the event 6 at time $t + x$. The time interval between the emitted pulses is related to the time interval between the received pulses as

$$t + x = k(t' + x').$$ (5.5)

Fig. 5.24 Observers A and B use the radar method to locate an object C. The observer A finds the object at (t, x)—event 4, while the observer B measures the same event at (t', x')

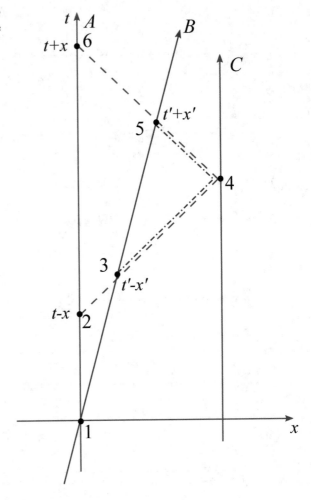

Equations (5.4) and (5.5) show the connection between the spacetime coordinates (t, x) and (t', x') for the same event 4, as measured by two different observers moving relative to each other.

In order to express the coordinates (t', x') in terms of the coordinates (t, x), we can first rewrite the Eqs. (5.4) and (5.5), opening the parentheses:

$$t' - x' = kt - kx,$$

$$kt' + kx' = t + x.$$

Multiplying both sides of the first equation by the k-factor and adding to the second equation, we get

$$2kt' = k^2t - k^2x + t + x, \tag{5.6}$$

from which follows

$$t' = \frac{k^2+1}{2k}t - \frac{k^2-1}{2k}x.$$

If, after multiplying both sides of the first equation by the k-factor, we subtract it from the second equation, we will obtain

$$2kx' = t + x - k^2t + k^2x,$$

and then get

$$x' = \frac{k^2+1}{2k}x - \frac{k^2-1}{2k}t.$$

It is a simple exercise to show that

$$\frac{k^2+1}{2k} = \frac{1}{\sqrt{1-v^2}} = \gamma, \tag{5.7}$$

and

$$\frac{k^2-1}{2k} = \frac{v}{\sqrt{1-v^2}} = \gamma v. \tag{5.8}$$

Finally, the relations between (t, x) and (t', x') become

▷ **Lorentz Transformation**

$$t' = \gamma(t - vx),$$
$$x' = \gamma(x - vt).$$

This set of equations is called the **Lorentz transformation**. They constitute the mathematical cornerstone of the special theory of relativity. All effects described so far, and more, can be deduced from these formulas. Although the Lorentz transformation allows one to study the special theory of relativity purely with algebra, we will continue to compliment the analyses of problems with Minkowski diagrams.

? **Problem 5.17**

Express (t, x) in terms of (t', x'). Hint: start with the Eqs. (5.4) and (5.5) and use steps similar to the ones used to find (t', x') in terms of (t, x).

? **Problem 5.18**

Write Lorentz transformation using the meter-based quantities $\bar{t}, \bar{x}$ and $\bar{v}$.

At small relative speeds the value of γ is very close to 1, the Lorentz transformation then reduces to *Voigt transformation* (after Woldemar Voigt, who derived them in 1887):

$$t' = t - vx,$$

$$x' = x - vt.$$

The second of these equations is familiar from the Galilean transformation. It describes the difference in the coordinates due to the increasing distance $d = vt$ between the observers. The first equation contains a new term: $vx = \bar{v}\bar{x}/c^2$. For the speeds and distances found on earth this term is very small, it is negligible in Galileo-Newtonian mechanics. However, it becomes significant even at small relative speeds if the distance x is large enough. For example, two observers in relative motion with the speed $v = 0.0001$ (the approximate speed of the Earth due to its orbital motion around the Sun), when measuring the time interval between the events on the opposite side of the Solar system ($x = 30$), will "disagree" by the amount of 0.003 s. This is not a vanishingly small amount for time measurements in our astronomical home.

Finally, for small relative speeds *and* small distances, Lorentz transformation reduces to Galilean transformation:

$$t' = t,$$

$$x' = x - vt.$$

Chapter Highlights

- Speed of electromagnetic interaction equals to the speed of other fundamental interactions (gravitational, weak,[8] and strong interactions) and represents and

---

[8] Electromagnetic and weak interactions are, according to the modern physics, parts of a single *electroweak interaction*.

invariant speed of causation. The invariance of the speed of light is used as a postulate in Einstein's special theory of relativity.

- Invariant speed of light c can be used to introduce c-based units, where all speeds are given as fractions of the speed of light, and all distances are given as time measured using the radar method.
- Minkowski diagrams are spacetime diagrams with both axes having the same units. In this book the units are seconds. Minkowski diagram is drawn for a specific reference frame. The worldlines of electromagnetic signals are diagonal in every Minkowski diagram, reflecting the fact that the speed of electromagnetic signal is the same for all frames of reference.
- The concept of simultaneity is relative: two events simultaneous in a reference frame S, are not simultaneous in any other reference frame S' in relative motion to S.
- The relative motion of two observers can be characterized by the rate of change of their relative distance. Another approach relies on the change of time interval between two successive electromagnetic signals. The relevant quantity is called Doppler k-factor. It is an important quantity in astrophysics.
- Velocity is a relative concept in the special theory of relativity. The transformation of velocity of an object between different reference frames is given by a *velocity composition* formula, which differs from simple addition.
- Clocks in relative motion can not remain synchronized. This fact follows from the path dependent nature of time measurements. Clocks measure the lengths of their worldlines—paths in spacetime. Two identical clocks "traveling" between a pair of events via different worldlines, will measure different total time. This is an absolute (invariant) fact, true for all reference frames.
- Lorentz transformation is the relationship between spacetime coordinates (t, x) and (t', x') of the same event, as measured by different observers in relative motion. Lorentz transformation contains Galilean transformation as the special case of small relative velocities of the observers *and* small distances between the events.

References

1. Einstein, A. *Albert Einstein: A Documentary Biography*, by Carl Seelig.
2. Minkowski, H. (1952). *Space and Time, in the Principle of Relativity*. New York: Dover.
3. Frisch, D. H., & Smith, J. H. (1963). Measurement of the relativistic time dilation using μ-mesons. *American Journal of Physics, 31*, 342–355.
4. Feynman, R., Leighton, R.B., & Sands, M. (2008). *Feynman Lectures on Physics* (vol. VIII). Section 18-3.
5. Taylor, E., & Wheeler, A. (1992). *Spacetime Physics* (2nd ed., p. 78). New York: Freeman and Company.

Chapter 6
Special Relativity B

Henceforth space by itself, and time by itself, are doomed to fade away into mere shadows, and only a kind of union of the two will preserve an independent reality.

H. Minkowski, *"Space and Time"*.

Abstract Using the Lorentz transformation we can deduce all effects of the special theory of relativity. In this chapter, we will first revisit already familiar results, such as the relativity of simultaneity and composition of velocities. Then, we will derive new effects, relatedal to the order of events and length measurements.

6.1 Simultaneity

Statements like "two events are simultaneous" or "two events happen at the same time" are not absolute—being true in one reference frame S, they become false in any other frame S' in a uniform and rectilinear motion relative to S. This follows from Lorentz transformation, as we now demonstrate.

Consider two events, E_1 and E_2, measured by the observers A and B in relative motion with speed v. The first event E_1 is measured by the observer A to happen at (t_1, x_1), and the second event E_2 is measured to happen at (t_2, x_2). The spacetime coordinates of these events, as measured by the observer B, can be found from the Lorentz transformation

$$t_1' = \gamma(t_1 - vx_1),$$
$$x_1' = \gamma(x_1 - vt_1), \tag{6.1}$$

© The Author(s), under exclusive license to Springer Nature Switzerland AG 2022
Y. Deshko, *Special Relativity*, Undergraduate Lecture Notes in Physics,
https://doi.org/10.1007/978-3-030-91142-3_6

for the event E_1, and

$$t_2' = \gamma(t_2 - vx_2),$$
$$x_2' = \gamma(x_2 - vt_2), \tag{6.2}$$

for the event E_2.

According to the observer A, the time interval between the events E_1 and E_2 is $\Delta t = t_2 - t_1$, while the observer B measures the time interval

$$\Delta t' = t_2' - t_1' = \gamma(\Delta t - v\Delta x), \tag{6.3}$$

where $\Delta x = x_2 - x_1$.

If the observer A finds that $\Delta t = 0$, meaning that the events happen at the same time, the observer B, in general, finds non-zero time interval

$$\Delta t' = -\gamma v \Delta x.$$

The events will be simultaneous for B only if either $\Delta x = 0$ or $v = 0$. When $\Delta x = 0$, we have two events happening at the same time in the same place, which is effectively a single event. The second condition, $v = 0$, means that the observers A and B are at rest relative to each other. In all other cases the events *will not be simultaneous*.

6.2   Order of Events

Not only simultaneity is relative, but even the ordering in time of *some* events can be different for different observers. Given two events, E_1 and E_2, with spacetime coordinates (t_1, x_1) and (t_2, x_2) respectively, their order in time is determined by the sign of time interval defined as

$$\Delta t = t_2 - t_1.$$

If $\Delta t > 0$ then the event E_2 happens *after* the event E_1. Negative Δt means that the event E_2 happens *before* the event E_1.

The time between the events E_1 and E_2 as measured by the observer B is

$$\Delta t' = \gamma(\Delta t - v\Delta x) = \gamma \Delta t \left(1 - v\frac{\Delta x}{\Delta t}\right).$$

The order of events will be different for the observer B if the sign of $\Delta t'$ will be different from Δt; mathematically this requirement can be written as follows:

$$\frac{\Delta t'}{\Delta t} < 0.$$

This inequality is satisfied when

$$1 - v\frac{\Delta x}{\Delta t} < 0 \quad \rightarrow \quad \Delta x > \frac{\Delta t}{v}. \tag{6.4}$$

For a given time between the events Δt, and the relative speed v of the observers A and B, if two events are separated by a sufficient distance Δx, their order in time will be different for A and for B: The observer A will measure E_2 as happening after E_1, while B will measure it as happening before, and vice versa.

> **Example:**
> In the reference frame of A a pair of events is separated by $\Delta t = 1$s, and $\Delta x = 4$s. For an observer B, moving with the speed $v = 0.8$ ($\gamma = 1.6$) relative to A, the time interval between the same events equals
>
> $$\Delta t' = 1.67 \times (1 - 0.8 \times 4) = -3.67\,s.$$
>
> The same events have different order in time for B. In contrast, if the events happen closer, e.g., $\Delta x = 0.4$s, then
>
> $$\Delta t' = 1.67 \times (1 - 0.8 \times 0.4) = 1.13\,s.$$
>
> The events will have the same order for B and for A.

Another possibility for a pair of events to have different order for different observers arises when the relative speed is sufficiently high. For a given time Δt and distance Δx between the events, if the observer B is moving with

$$v > \frac{\Delta t}{\Delta x},$$

then it will measure the events E_1 and E_2 happening in the order opposite to what A measures.

> **Example:**
> In the reference frame of A a pair of events is separated by $\Delta t = 1$s, and $\Delta x = 4$s. For an observer B, moving with the speed $v = 0.1$ ($\gamma = 1.005$)

(continued)

relative to A, the time interval between the same events equals

$$\Delta t' = 1.005 \times (1 - 0.1 \times 4) = 0.603\,s.$$

The events have the same order in time for B. In contrast, if the relative speed equals $v = 0.5$ ($\gamma = 1.15$), the time interval becomes

$$\Delta t' = 1.15 \times (1 - 0.5 \times 4) = -1.15\,s.$$

For B, the events will have the order opposite to the order for A.

Not every pair of events can have their order reversed, but only those that are not *causally connected*. It would be absurd to see the light from a flash before the flash happens. If causation requires a signal of some sort to propagate from an event E_1 (cause) to an event E_2 (effect), then the time between these events will be

$$\Delta t = \frac{\Delta x}{u_c},$$

where u_c is the speed of the causation signal (relative to A, of course). As we showed, the cause and effect events may change their order if

$$\Delta x > \frac{\Delta t}{v} \quad \rightarrow \quad \Delta x > \frac{\Delta x}{v u_c}. \qquad (6.5)$$

From the last equation we find the requirement for the relative speed:

$$1 > \frac{1}{v u_c}.$$

In other words, the observer B must be moving relative to A with the speed

$$v > \frac{1}{u_c}. \qquad (6.6)$$

? Problem 6.1

Imagine that every cause creates its effect instantly in a given reference frame A. This means that the speed of causation is infinite

$$u_c = \infty.$$

What does this mean for the order of events, as measured by different observers in motion relative to A?

In all known fundamental interactions, such as electromagnetic,[1] gravitational, and strong, the signal responsible for the interaction propagates with the same speed of approximately 300,000,000 meters per second. It is known as the speed of electromagnetic waves in vacuum, but just as well could be associated with the speed of gravitational waves, or the speed of the "waves" of strong nuclear interaction. The important fact is this:

▷ **Universal Speed of Causation**

There exists an absolute speed of causation signal. It is equal to the speed of electromagnetic waves, gravitational waves, and the speed of waves of other fundamental interactions.

Given that the causation speed is $u_C = 1$, the speed of the observer B, required to reverse the order of two causally connected events, must be

$$v > 1.$$

The observer must be traveling faster than the signal that connects the cause and effect. Faster than light.

For additional information about absolute speed of causation, see the Appendix B.4.

$$* * *$$

We can summarize the properties of time measurements, discovered so far, as follows:

▷ **Properties of Time Measurements**

- *Clocks in relative motion do not remain synchronized.*
- *Time elapsed on a clock depends on the path the clock takes "through" spacetime.*
- *Time between two events is not invariant. Events can not be simultaneous for all inertial observers.*

---

[1] Strictly speaking, electromagnetic is the part of so called electroweak interaction, but this is not very important here.

• *Events that are not causally connected have no absolutely fixed order in time.*
  Their order may be different for different observers.

? Problem 6.2

The distance between the Milky Way galaxy and the Andromeda galaxy is 2,573,000 light-years. An observer A measures two events happening in these galaxies as simultaneous. What is the time interval between these events as measured by an observer B, moving with the speed $v = 0.0001$ relative to A.

6.3   Velocity Composition

Starting with Galilean transformation we derived, in Sect. 4.1.5, the formula connecting the velocity of an object as measured by different observers in relative motion. Using similar arguments as before, we can find the formula of velocity composition using Lorentz transformation. This is an algebraic solution to the problem solved using Minkowski diagrams in the Sect. 5.9.

Consider an observer A locating an object X at time t_1 to be at the position x_1 (event E_1); and, at a later time t_2, at the position x_2 (event E_2). The velocity of the object X is measured by A as

$$u = \frac{x_2 - x_1}{t_2 - t_1} = \frac{\Delta x}{\Delta t}.$$

An observer B, moving with the velocity v relative to A, assigns the following spacetime coordinates to the events E_1 and E_2:

$$t_1' = \gamma(t_1 - vx_1),$$

$$x_1' = \gamma(x_1 - vt_1),$$

and

$$t_2' = \gamma(t_2 - vx_2),$$

$$x_2' = \gamma(x_2 - vt_2).$$

From these equations follows that

$$\Delta t' = t_2' - t_1' = \gamma(\Delta t - v\Delta x),$$

$$\Delta x' = x_2' - x_1' = \gamma(\Delta x - v\Delta t).$$

According to the observer B, the velocity of the object X is

$$u' = \frac{\Delta x'}{\Delta t'} = \frac{\Delta x - v\Delta t}{\Delta t - v\Delta x}.$$

Using the fact that $\Delta x = u\Delta t$, we obtain

$$u' = \frac{u - v}{1 - vu}.$$

This formula differs from (5.3), obtained using the Minkowski diagram. The reason is that now we express the velocity of the object relative to B in terms of the velocity relative to A. The difference is in the sign of the relative velocity of the observers. It is a simple algebraic exercise to solve this equation for u:

$$u = \frac{u' + v}{1 + vu'}.$$

6.3.1 Fizeau Experiment

The rule for the velocity composition provides a relativistic explanation of the results of an important optical experiment, performed in 1851 by a French physicist Hippolyte Fizeau. The experiment was repeated in a more refined manner by Michelson and Morley in 1886, one year before their another, more famous experiment. The idea of the experiment is illustrated in Fig. 6.1.

(a)

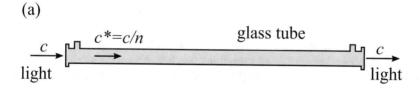

(b)

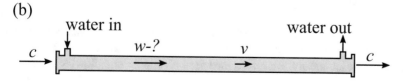

Fig. 6.1 (a) The speed of light in water c^* is less than in vacuum: $c^* = c/n < 1$. (b) Water running through a tube "carries" light with it, making the speed of light in running water greater than the speed of light in water at rest, when measured relative to the laboratory

When light enters a transparent medium, such as glass or water, it slows down. The slowing down is due to the interaction of light and the atoms or molecules of the medium. The interaction processes are quite complex, but the net effect is that the speed of light through a transparent medium is given by

$$c^* = \frac{c}{n},$$

where the number n is called *index of refraction*; it is characteristic to a given medium, for example, $n_{glass} \approx 1.4$ and $n_{water} \approx 1.3$.

In the experiment performed by Fizeau, the light was sent through a tube filled with water, as illustrated in Fig. 6.1. Both sides of the tube were transparent, to allow the light to enter and exit. The tube also allowed water to enter and exit at the opposite sides, flowing through the tube with some fixed speed v.

The tube shown in Fig. 6.1 was a part of a sensitive optical instrument capable of detecting tiny changes in the speed of light relative to the tube. Having performed the optical measurements with the water at rest relative to the tube, and then with the water running with the speed v, Fizeau established that in the latter case the speed of light relative to the laboratory (and the tubes) is given by

$$c^*(v) = \frac{c}{n} + v \cdot \omega, \tag{6.7}$$

where the coefficient

$$\omega = 1 - \left(\frac{c}{n}\right)^2$$

was called the *drag coefficient*. Its quantifies how strongly the flowing water affects the propagation of light trough water. For $\omega = 1$ the water would completely "drag" the light in it, whereas for $\omega < 1$ the "drag" would be partial.

The explanation of this result in the times of Fizeau was based on the idea of *ether* (or aether)—a physical medium responsible for the propagation of light. It was assumed that the ether permeated everything and was present between the molecules of water. When water flowed through the tube, it could "drag" the ether with it, thus allowing light to move faster relative to the tubes and the laboratory. However, the result (6.7) can be obtained using the velocity composition formula without referring to any additional assumptions like ether.

First, let us consider the case of flowing water in the reference frame S' co-moving with the water. In this reference frame the water is at rest and the speed of light through it is $u = c/n$. The reference frame S' is moving with the speed v relative to the laboratory. Therefore, the speed of light through flowing water relative to the laboratory is

$$w = \frac{v + u}{1 + uv} = \frac{v + c/n}{1 + cv/n}.$$

The speed of water is small compared to the speed of light, so we can use the only the first two terms in the geometric series expansion[2]

$$\frac{1}{1 + cv/n} \approx 1 - \frac{cv}{n}$$

and write

$$w \approx \left(v + \frac{c}{n}\right)\left(1 - \frac{cv}{n}\right).$$

Opening the parentheses and keeping only the terms proportional to the first power of v, we obtain

$$w \approx \frac{c}{n} + v\left[1 - \left(\frac{c}{n}\right)^2\right].$$

This is the formula experimentally established by Hippolyte Fizeau and Michelson and Morley.

6.4 Length Measurement

As we established, simultaneity is relative: observers in relative motion will not agree whether two events happen "at the same time." This fact has important consequences for the measurement of distances. Consider, for example, the situations when we need to determine any of the following:

1. A length L of an object.
2. Coulomb force between two charges:

$$F = k\frac{Q_1 Q_2}{r^2}.$$

3. Newtonian force of gravity between two asteroids:

$$F = G\frac{M_1 M_2}{r^2}.$$

4. Force of a stretched spring on the attached object:

$$F = H(x - x_0).$$

---

[2] See Appendix A.1.

In all these cases, we need to determine positions of two objects (or points) located in different places. When the objects in question are moving relative to the observer, it becomes necessary to locate them *at the same time*, in order to determine the distance properly.

Consider two observers—A and B—moving relative to each other with the speed v. The observer B has a stick at rest in its reference frame. Let both observers measure the positions of the left and the right edges of the stick. We may assume that the radar method is used to shine a light on mirrors attached to each end of the stick. Let the event E_1 be the reflection of the light from the left end of the stick, and the event E_2—the reflection of the light from the right end of the stick. If the observer A assigns the spacetime coordinates (t_1, x_1) and (t_2, x_2) to these events, the observer B will have

$$t_1' = \gamma(t_1 - vx_1),$$
$$x_1' = \gamma(x_1 - vt_1),$$

and

$$t_2' = \gamma(t_2 - vx_2),$$
$$x_2' = \gamma(x_2 - vt_2),$$

for the events E_1 and E_2, respectively.

For a moving object it is important to locate the ends at the same time, otherwise the object will move in between the measurements. Therefore, the observer A requires that $t_2 = t_1$ and determines

$$L = x_2 - x_1$$

as the length of the moving stick. As we showed earlier, the observer B will not find the events E_1 and E_2 simultaneous, but it does not matter, since for a stationary object we can locate the left end on Monday and the right one on Friday. Thus, for the observer B the events E_1 and E_2 can also be used for the measurement of the length of the stick. The distance the observer B will measure is

$$L_0 = x_2' - x_1' = \gamma(x_2 - x_1) - \gamma v(t_2 - t_1),$$

and since $t_1 = t_2$, we obtain

$$L_0 = \gamma(x_2 - x_1) = \gamma L.$$

The length of the moving object, as measured by A, is then given by

$$L = L_0\sqrt{1 - v^2}. \tag{6.8}$$

The results of these straightforward measurements are different! The observer A measures the moving stick to be shorter than the same stick at rest.

? Problem 6.3

Two rulers with equal lengths at rest L_0 are moving relative to some reference frame S with the speeds $v = 0.3$ and $v = 0.8$, respectively. Draw Minkowski diagram of these rulers, as measured in S.

Another useful view of this problem is provided by a Minkowski diagram comparing the events that happen at (t, L_0) for the observer A, to the events happening at (t', L_0) for the observer B. These events indicate the worldlines of the right edges of two identical sticks, one at rest relative to A, the other at rest relative to B.

The events with spacetime coordinates (t', L_0) relative to the observer B will receive spacetime coordinates

$$t = \gamma(t' + vL_0) = \gamma t' + \gamma v L_0,$$

$$x = \gamma(L_0 + vt') = \gamma L_0 + \gamma v t',$$

in A's reference frame. From the second equation we find

$$\gamma t' = \frac{x - \gamma L_0}{v}$$

and plug it into the first equation to get "t vs. x":

$$t = \frac{x}{v} - \frac{L_0}{\gamma v}.$$

The worldline described by this equation is a straight line with the same slope as the worldline of the observer B (the observer B is at the origin of its coordinate system, coinciding with $L_0 = 0$). It intersects the line $t = 0$ at $x_0 = L_0/\gamma$, which corresponds to the length of the B's stick, as measured by A at the moment $t = 0$. Minkowski diagram in Fig. 6.2 illustrates this point.

Different observers, moving relative to the stick with different speeds, will measure different lengths. Since there are infinite number of equally valid observers, there will be infinite number of measured lengths of a moving object. Which one, among this infinity, is the "real length", *if there is such a thing*? To answer this question, we must understand how different observers in relative motion "slice" the same spacetime into simultaneous events.

For the observer A all events that happen at the same time $t = T$ are indicated by a horizontal line parallel to the x axis, see Fig. 6.3 These events will not be

Fig. 6.2 Identical sticks in
relative motion will have
different measured lengths.
The stick with the length L_0
at rest relative to the observer
B will be shorter than
identical stick of length L_0 at
rest relative to A

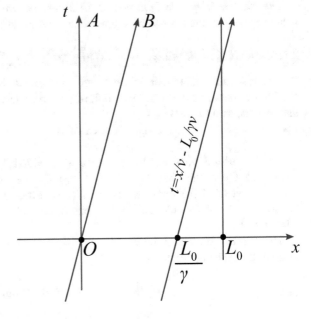

Fig. 6.3 Observers in
relative motion "slice"
spacetime into lines of
simultaneity differently. For
the observer A all events that
happen at the same time T
are indicated by the
horizontal red dotted line.
The events that happen at the
same time T, *according to the
observer B*, are indicated by
a sloped blue dashed-dotted
line. This line intersects the t
axis at $(T/\gamma, 0)$

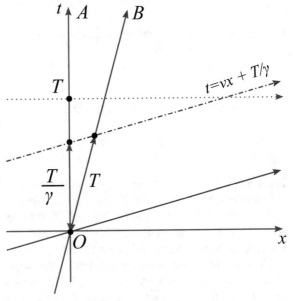

simultaneous for the observer B, moving relative to A. According to the observer
B, the events that happen at the time T of its clock have spacetime coordinates
(T, x'). In Minkowski diagram for the reference frame A, these events will have
coordinates

Fig. 6.4 A stick, or a ruler, or any other extended object is a *whole spacetime configuration of events*, shown here as the shaded region. It is not a spatial "slice" of this configuration, but the totality of related events. Observers in relative motion "slice" the shaded spacetime region with lines of simultaneity differently, thus measuring different lengths. *None of those lengths is the "true" length—they are all equivalent*

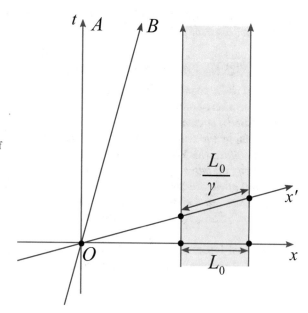

$$t = \gamma(T + vx') = \gamma T + \gamma vx',$$

$$x = \gamma(x' + vT) = \gamma x' + \gamma vT.$$

From the last equation we find

$$\gamma x' = x - \gamma vT$$

and plug it into the first equation to obtain "t vs. x":

$$t = vx + T/\gamma.$$

This equation describes a straight line in the xt plane of Minkowski diagram. The slope of the line is proportional to the relative speed v. It intersects the time axis at $x = 0$ and $t = T/\gamma$, as illustrated in Fig. 6.3.

We can now understand the difference in the length measurements, performed by different observers in relative motion, in the following way. A stick, or any other extended object, is made of material points whose worldlines span a certain region in spacetime. Figure 6.4 shows one such region for an object of the length L_0, at rest relative to A.

To measure the length of a moving object, an observer must select events that happen at the same time. This results in "slicing" the spacetime strip differently for different observers, as the diagram in Fig. 6.4 demonstrates. All such "slicings" are equally valid; in other words, *all measured lengths are equally "true"*. Max Born, in his book *"Einstein's Theory of Relativity"* expressed this most lucidly:[3]

▷ Max Born on Length

A material rod is physically not a spatial thing but a space-time configuration[...]; it is not a section of the x-axis but rather a strip of the x, ct plane.

If we slice a cucumber, the slices will be larger the more obliquely we cut them. It is meaningless to call the sizes of the various oblique slices "apparent" and call, say, the smallest which we get by slicing perpendicularly to the axis the "real" size.

In exactly the same way a rod in Einstein's theory has various lengths according to the point of view of the observer. One of these lengths, the static or proper length, is the greatest, but this does not make it more real than the others.

Additional information about the notion of objects and their lengths, from the standpoint of the special theory of relativity, is given in the Appendix B.5.

6.4.1   FitzGerald and Lorentz Hypotheses

The length of an object moving with the speed v relative to an observer is

$$L = L_0\sqrt{1 - v^2},$$

where L_0 is the length of this object at rest. The same formula was derived by Lorentz in 1895, building on the ideas that he and FitzGerald put forward around 1890. Both Lorentz and FitzGerald thought that light propagated in *ether* and that the motion of material objects relative to the ether was responsible for certain physical phenomena that happen to the moving objects.

Both FitzGerald and Lorentz were motivated by the results of the 1887 experiment of Michelson and Morley. The details of the experiment are not important here. In short, Michelson and Morley, using a very sensitive optical method, demonstrated the absence of "ether wind", despite the expected motion of the Earth through it. FitzGerald and Lorentz proposed that the "ether wind" affected the measurement apparatus of Michelson and Morley, rendering their method ineffective.

FitzGerald suggested the following:[4]

---

[3] Max Born, *Einstein's Theory of Relativity*, Dover, 1965, pp. 253–255.
[4] FitzGerald, *The Ether and the Earth's Atmosphere*, Science 17 May 1889, Vol. 13, Issue 328, pp. 390.

> ## ▷ FitzGerald Hypothesis

... the length of material bodies changes, according as they are moving through the ether or across it, by an amount depending on the square of the ratio of their velocity to that of light. We know that electric forces are affected by the motion of the electrified bodies relative to the ether, and it seems a not improbable supposition that the molecular forces are affected by the motion, and that the size of a body alters consequently.

A similar hypothesis was advanced by Lorentz:[5]

> ## ▷ Lorentz Hypothesis

Now we can assume at present, that electric and magnetic forces act by intervention of the aether. It is not unnatural to assume the same for molecular forces, but then it can make a difference, whether the connecting line of two particles, which move together through the ether, is moving parallel to the direction of motion or perpendicular to it.

... If therefore S_2 is the state of equilibrium of a solid body at rest, then the molecules in S_1 have precisely those positions in which they can persist under the influence of translation. The displacement would naturally bring about this disposition of the molecules of its own accord, and thus effect shortening in the direction of motion in the proportion of 1 to $\sqrt{1 - v^2/c^2}$.

The FitzGerald-Lorentz contraction hypothesis *is very different from the effect of motion on length measurement we discussed above*. The two *must not be confused*, even though formulas look the same. The idea of FitzGerald-Lorentz is illustrated in Fig. 6.5.

Firstly, the FitzGerald-Lorentz contraction was hypothesized for solid object held by electric and molecular forces. The distance between two planets moving through the ether would remain the same. This is not so in Einstein's special theory of relativity; see Fig. 6.5c, d.

Secondly, the change of the length of a solid object is *absolute*—happens for all observers, including the observer at rest relative to the object. Figure 6.6 shows a configuration with the solid rod and two mirrors near its edges freely floating in space. Importantly, the mirrors are *not* attached to the rod, they are simply aligned. According to the FitzGerald-Lorentz contraction hypothesis, all solid objects contract if they are moving through the ether. In the reference frame where the rod is at rest, the ether is moving relative to the rod, causing its contraction. The thickness of each mirror is reduced as well, but the distance between the mirrors

---

[5] H. A. Lorentz, *The Relative Motion of the Earth and the Aether*, 1892.

Fig. 6.5 (**a**) A rod at rest relative to a hypothetical ether; the ether is schematically shown as scattered blue dots. (**b**) The same rod, moving relative to the ether, experiences an "ether wind". According to the FitzGerald-Lorentz hypothesis, the forces between the molecules of the rod change, resulting in the shortening of the rod in the direction of motion. To keep the figure clean, only several "ether particles" are shown as moving relative to the rod. (**c**), (**d**) In contrast to the special theory of relativity, the distance between two planets moving through the ether would remain unchanged, according to FitzGerald-Lorentz hypothesis; the planets, however, will be shortened

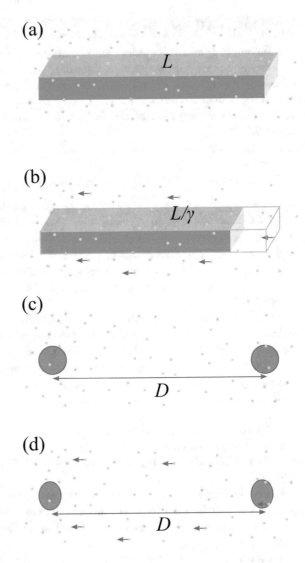

remains the same. This is different from the prediction of the special theory of relativity.

Everything is Relative?

A superficial familiarity with the special theory of relativity may create an impression that everything we knew about physics is turned upside down: Velocity, simultaneity, order of events, lengths are different for different observers. "Everything is relative", one may think and use the authority of Albert Einstein to support this view.

Fig. 6.6 (a) A rod at rest relative to the ether; two mirrors are floating freely near the edges of the rod. Whole construction is in space. (b) The same rod and the mirrors moving relative to the ether. In the reference frame of the rod the "ether wind" causes solid bodies to contract, including the rod and each mirror separately. However, the region between the mirrors is empty space and the distance must remain the same. This is contrary to the special theory of relativity

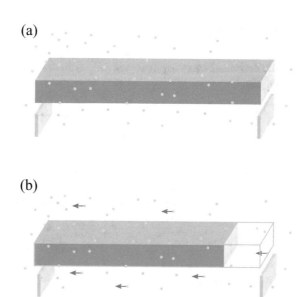

In an interview given to George Sylvester Viereck,[6] Albert Einstein commented:

▷ **Einstein On Relativity**

The meaning of relativity has been widely misunderstood. Philosophers play with the word, like a child with a doll. Relativity, as I see it, merely denotes that certain physical and mechanical facts, which have been regarded as positive and permanent, are relative with regard to certain other facts in the sphere of physics and mechanics. It does not mean that everything in life is relative and that we have the right to turn the whole world mischievously topsy-turvy.

To be fair, modern philosophers, and especially the philosophers of science, are much more aware of the proper interpretation of the relativity theory.

The special theory of relativity reveals that certain physical facts, such as simultaneity, are not absolute, in contrast to common-sense expectations based on every-day experience. But in addition to identifying relative quantities, the theory points to *absolute*, or *invariant*, quantities of a new kind. The first absolute quantity we will find is based on the combination of time interval and distance between two events.

---

[6] G. S. Viereck *What Life Means to Einstein*, The Saturday Evening Post, October 26, 1929.

6.5   Fundamental Invariant

While deriving Lorentz transformation, in the Sect. 5.13, we obtained the following equations:

$$(t' - x') = k(t - x),$$
$$k(t' + x') = (t + x). \tag{6.9}$$

These equations relate the position and time of the same event measured by two observers in relative motion.

The product of the left-hand sides must be equal to the product of the right-hand sides:

$$k(t' + x')(t' - x') = k(t + x)(t - x),$$

which simplifies to

$$t'^2 - x'^2 = t^2 - x^2.$$

We found an *invariant* or *absolute* quantity—the same for any two observers. This invariant combines the measurements of time and space coordinates.

? Problem 6.4

Use Lorentz transformation to show that for two events $E_1 = (t_1, x_1)$ and $E_2 = (t_2, x_2)$ the quantity

$$(t_2 - t_1)^2 - (x_2 - x_1)^2$$

is the same in all inertial reference frames.

We will call the invariant quantity Δs, defined as

$$\Delta s^2 \equiv (t_2 - t_1)^2 - (x_2 - x_1)^2 = \Delta t^2 - \Delta x^2 \tag{6.10}$$

a spacetime *separation* or spacetime *interval*. It is an important, fundamental invariant in the special theory of relativity.

Notice how the expression for the spacetime interval resembles the expression for the distance in planar Euclidean geometry

$$d^2 = x^2 + y^2.$$

Similar to the distance in Euclidean geometry, spacetime interval is invariant, although the invariance is with respect to Lorentz transformation (change of inertial reference frame). We might call it the distance in Minkowski diagram or *Minkowski distance*. There is, however, a significant difference between Euclidean distance and Minkowski distance: The distance squared d^2 is always positive, whereas the quantity

$$\Delta s^2 = \Delta t^2 - \Delta x^2$$

can be positive, negative, or zero.

? Problem 6.5

The lifetime of a muon at rest is T_0. Using the invariance of the interval between two events (e.g., the birth and decay of a muon)

$$(\Delta t')^2 - (\Delta x')^2 = (\Delta t)^2 - (\Delta x)^2,$$

find how fast the muon must travel relative to an observer, to cover the distance D.

6.5.1 Three Kinds of Interval

There are three kinds of spacetime interval Δs, depending on the sign of the expression

$$\Delta s^2 = \Delta t^2 - \Delta x^2.$$

Each kind of spacetime interval has a distinct and clear physical meaning. Let us start with positive intervals.

Consider an observer measuring the spacetime coordinates of two events, E_1 and E_2, that happen in the same location x but at different times t_1 and t_2, as shown in Fig. 6.7. In this case the separation reduces to

$$\Delta s^2 = (t_2 - t_1)^2 - (x - x)^2 = \Delta t^2 > 0. \tag{6.11}$$

The magnitude of Δs gives the time between the events E_1 and E_2. It can be measured by an inertial observer whose worldline passes through these events. Thus, a regular clock measures positive spacetime interval, which is called *time-like interval*.

Next, consider two events, $E_1 = (t, x_1)$ and $E_3 = (t, x_3)$, that happen at the same time t according to some observer A; see Fig. 6.7. The spacetime separation squared is

Fig. 6.7 Three kinds of
spacetime interval: positive
(E_1-E_2), negative (E_1-E_3),
and zero (E_1-E_4)

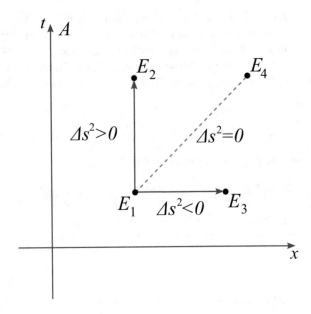

$$\Delta s^2 = \Delta t^2 - \Delta x^2 = -\Delta x^2 = -d^2, \tag{6.12}$$

which is simply a minus squared distance between two locations in space. A regular
ruler measures negative spacetime interval, which is called *space-like interval*.

Finally, when the interval between two events is zero, the slope of a worldline
passing through these events must be equal ± 1:

$$\Delta s^2 = \Delta t^2 - \Delta x^2 = 0 \;\rightarrow\; v = \frac{\Delta x}{\Delta t} = \pm 1. \tag{6.13}$$

In a spacetime diagram such a pair of events can be connected by a worldline of
light signal:[7] such spacetime interval is called *light-like*.

The examples considered above are, in a certain sense, special; they can be
generalized; see Fig. 6.8.

Firstly, events separated by a time-like interval do not have to happen in the
same place for a given observer, but they must happen in the same place for *some*
observer. To see this, consider an observer B moving relative to the observer A with
the speed v. The observer A locates the observer B at the events $E_1 = (t_1, x_1)$ and
$E_2 = (t_2, x_2)$, as shown in Fig. 6.8. The interval squared for this pair of events is

$$\Delta s^2 = (t_2 - t_1)^2 - (x_2 - x_1)^2 = \Delta t^2 - \Delta x^2. \tag{6.14}$$

---

[7] Or anything else traveling at the speed of light.

Fig. 6.8 More general cases
of time-like (E_1-E_2) and
space-like (E_1-E_3) intervals.
Note how a time-like interval
lies within the top quadrant,
defined by the worldlines of
light, whereas a space-like
interval lies outside of that
region. An observer may be
present at both E_1 and E_2; no
observer can be present at
both E_1 and E_3

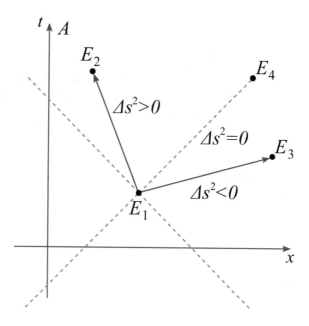

Since the observer B is moving with a constant speed v, we have $\Delta x = v\Delta t$, and
the last expression for the interval squared takes the form

$$\Delta s^2 = \Delta t^2 (1 - v^2) = \left(\Delta t / \gamma\right)^2 > 0. \qquad (6.15)$$

The spacetime coordinates of the same events, as measured by the observer B, are

$$t_1' = \gamma (t_1 - v x_1),$$
$$x_1' = 0, \qquad (6.16)$$

and

$$t_2' = \gamma (t_2 - v x_2),$$
$$x_2' = 0. \qquad (6.17)$$

The interval squared between these events for the observer B is given by

$$\Delta s^2 = \Delta t'^2 - \Delta x'^2 = \Delta t'^2 > 0. \qquad (6.18)$$

Thus, the interval between the events E_1 and E_2 is time-like and corresponds to
the time between these events *as measured by an inertial observer B present at both
events*:

$$\Delta s = \Delta t' = \frac{\Delta t}{\gamma}.$$

Secondly, two events $E_1 = (t_1, x_1)$ and $E_3 = (t_3, x_3)$ with space-like separation do not have to be simultaneous for an observer A, but they will be simultaneous for *some* observer B. Indeed, let the spacetime separation between E_1 and E_3 be space-like, as in Fig. 6.8:

$$\Delta s^2 = (t_3 - t_1)^2 - (x_3 - x_1)^2 < 0. \tag{6.19}$$

If the observer B is moving relative to A with the speed v, it will assign the following spacetime coordinates to the events E_1 and E_3:

$$t_1' = \gamma(t_1 - vx_1),$$
$$x_1' = \gamma(x_1 - vt_1);$$

and

$$t_3' = \gamma(t_3 - vx_3),$$
$$x_3' = \gamma(x_3 - vt_3).$$

The requirement of the simultaneity for the events E_1 and E_3 ($t_3' = t_1'$) leads to

$$t_3 - vx_3 = t_1 - vx_1,$$

and, consequently,

$$v = \frac{t_3 - t_1}{x_3 - x_1} = \frac{\Delta t}{\Delta x}.$$

(*Note*: This is not a mistake to have velocity equal to "time over distance"; this formula is *not* a definition, but simply the relation between the relative velocity of the observers A and B and the spacetime coordinates of the events E_1 and E_3.)

Since the interval squared between the events E_1 and E_3 is negative, we have

$$\Delta t^2 - \Delta x^2 < 0 \quad \rightarrow \quad \Delta t < \Delta x,$$

implying that the required speed v is less than the speed of light. Thus, a pair of space-like separated events is simultaneous for *some* observer. The interval corresponds to the distance between these events, as measured by this particular observer, which finds the events to be simultaneous.

Writing interval squared as

$$\Delta s^2 = (|\Delta t| - |\Delta x|)(|\Delta t| + |\Delta x|), \tag{6.20}$$

Table 6.1 To determine the type of the interval between a pair of events we can compare their separation in time with the separation in space. Another test is to see if an observer can travel between these events

Separation	Property	Comment
time-like	$|\Delta t| > |\Delta x|$	Some observer can travel between the events
light-like	$|\Delta t| = |\Delta x|$	Light can travel between the events
space-like	$|\Delta t| < |\Delta x|$	Nothing can travel between the events

we can see that the sign of the interval squared, and therefore the kind of the interval, is determined only by the first factor

$$|\Delta t| - |\Delta x|.$$

To determine the type of the interval between a pair of events, we need to compare their separation in time, along the t axis, to the separation in space, along the x axis. If the events are farther in time, compared to their separation in space, then the interval is time-like and can be measured by a clock; if the events are farther in space, compared to their separation in time, then the interval is space-like and can be measured by a ruler; if the events are equally separated in time and space then the interval is light-like and corresponds to the propagation of light between the events. These results are summarized in Table 6.1.

Labeling intervals as time-like or space-like is better than relying on the sign of the invariant interval squared Δs^2. In some works on the theory of relativity the interval squared is defined differently:

$$\Delta s^2 = \Delta x^2 - \Delta t^2,$$

and the convention based on the sign will be the opposite to what was defined earlier. Using the terms *time-like* and *space-like* helps avoid the ambiguity caused by the difference in the definition of the spacetime interval squared.

6.5.2 Distance in Spacetime

In Euclidean geometry distance between two points is an *invariant* quantity, it does not depend on the coordinate system used. The formula for the distance squared between two points looks the simplest in Cartesian coordinates:

$$d^2 = (x_2 - x_1)^2 + (y_2 - y_1)^2,$$

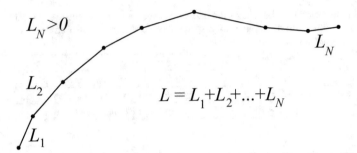

Fig. 6.9 In Euclidean geometry the distance squared between any two points is positive. The length of a curve is the sum of length of its parts

where (x_1, y_1) and (x_2, y_2) are the Cartesian coordinates of the first and the second points, respectively.

The coordinates *of the same points* will change when we choose another Cartesian coordinate system, but the form of the distance squared, as well as its value, remains the same:

$$d^2 = (x_2' - x_1')^2 + (y_2' - y_1')^2.$$

In Euclidean geometry the distance squared between any two distinct points is always a positive quantity. The length of a path in a plane can be found as the sum of lengths of separate pieces making up the path, as shown in Fig. 6.9.

In the special theory of relativity there is also an invariant quantity—the interval between two events. The events receive spacetime coordinates once a reference frame is chosen. *The analogue of a Cartesian coordinate system for spaceime is an inertial reference frame.* In an inertial reference frame with Cartesian coordinates the interval squared is given by

$$\Delta s^2 = (t_2 - t_1)^2 - (x_2 - x_1)^2.$$

In any other inertial reference frame with Cartesian coordinates, the value and the form of the interval squared remains the same

$$\Delta s^2 = (t_2' - t_1')^2 - (x_2' - x_1')^2.$$

The interval, therefore, can be viewed as the *distance in spacetime*. Unlike the Euclidean distance, spacetime distance has three "flavors": *time-like, light-like,* and *space-like.* Each kind of the interval expresses the relationship between two events that the interval connects.

Consider all events lying on a circle with "unit radius" in Minkowski diagram, as shown in Fig. 6.10. The interval squared between the events OA is positive and equals to the time between these events. Similarly, the interval squared between the

Fig. 6.10 In spacetime the
distance squared between a
pair of events can be either
positive (OB), negative
(OD), or zero (OC)

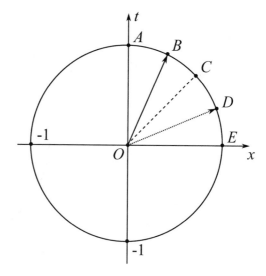

events OB is positive and equals to the time between these events, measured by an
observer whose worldline "goes through" these events. Notably,

$$\Delta s^2_{OA} > \Delta s^2_{OB}.$$

To see this, first write the equation for the "circle of events" in Minkowski
diagram:

$$t^2 + x^2 = 1,$$

then use this equation to find the interval squared

$$\Delta s^2 = t^2 - x^2 = 2t^2 - 1.$$

As the point on the circle moves away from the vertical axis t, the value of
the coordinate t decreases, reducing the value of Δs^2.

As B approaches the event C, the distance Δs_{OB} is getting smaller; it becomes
zero for the pair of events OC. The spacetime distance between the events OA is 1,
while for the events OC it is zero.

The interval squared between the events OE is negative. It is related to the
distance between these events:

Fig. 6.11 A curve in
spacetime can be made of a
mix of intervals: (**a**) time-like,
(**b**) light-like, and (**c**)
space-like. *There is no proper
way to add their length*

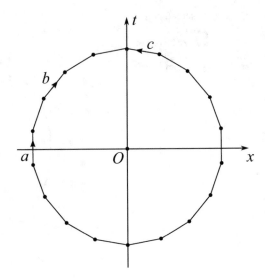

$$D_{OE} = \sqrt{-\Delta s_{OE}^2}.$$

The interval squared between the events OD is also negative, and corresponds to
the distance D_{OD}, measured by some observer which finds the events O and D
simultaneous. The distance D_{OD} is shorter than D_{OE}. As the event D is getting
closer to C, the distance D_{OD} is getting shorter, reaching zero in the limit $D \to C$.

Given that in spacetime the distance can be positive, negative, and even zero,
how do we find the length of an arbitrary curve? For example, the curve in
Fig. 6.11 is made of many "pieces", some are time-like, some are space-like, and
others may be light-like. It is mathematically possible to add up all times and
distances, corresponding to different parts of the curve; the result may be declared
the "length" of the curve. However, this approach is *not* used in the theory of
relativity. The reason is simple: *An arbitrary curve in spacetime has no clear
physical interpretation.*

An arbitrary motion of a particle, moving relative to an inertial observer, is
represented in Minkowski diagram by a curve (*worldline*) that is made of only time-
like "pieces". An example is given in Fig. 6.12, curve (a). The length of such a curve
between two events, for example 1 and 2, corresponds to the total time that would
elapse on a clock moving along with the particle.[8]

The curve (b) in Fig. 6.12 is the example of a space-like curve: each "piece"
of it is a space-like interval. Such a curve *does not correspond to any worldline.*
Furthermore, there exists no inertial observer that would measure all events from

---

[8] To rigorously prove this, we must know how accelerated motion affects time measurements. This
is the domain of general theory of relativity. Alternatively, we could break a smooth curve into
"inertial" pieces and use the ideas of infinite sums from calculus.

Fig. 6.12 In Minkowski diagrams there are time-like curves (**a**) and space-like (**b**) curves. Propagation of light is represented by straight lines and not shown here

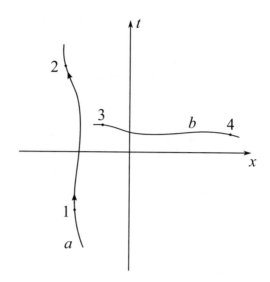

this curve as simultaneous. The sum of the lengths of all "pieces" of this curve does not correspond to any distance relative to an inertial observer. In the special theory of relativity, it is meaningless to speak of the distance between any two events, such as 3 and 4, measured along such curve.

▷ Distances in Spacetime

A spacetime distance can be measured along any time-like curve. This distance corresponds to the total time that would have elapsed on a clock "moving along" this curve.

? Problem 6.6

Draw Minkowski diagram and choose two events separated by a time-like interval.

(a) Draw the longest worldline between the events.
(b) Draw the shortest worldline between the events. (Hint: It does not have to be a straight line.) How many shortest paths are possible?

? Problem 6.7

In Euclidean geometry a circle is defined as all points equidistant from a specified center-point O. Given that in spacetime there are three types of distances, depending on the sign of the interval squared, one can draw three types of "spacetime circles".

(A) In Minkowski diagram for an observer A, draw curves that contain all events at constant spacetime distance from the origin, such that

$$t^2 - x^2 = 1.$$

(B) Draw curves that contain all events at constant spacetime distance from the origin, such that

$$t^2 - x^2 = -1.$$

(C) Draw curves that contain all events at constant spacetime distance from the origin, such that

$$t^2 - x^2 = 0.$$

6.6   Past, Future, and Present

The common sense view on the past, future, and present can be expressed as follows: Despite the difference of times at the opposite sides of the globe, there is a clear sense of a single *now* in New York and Bejing, as well as on the Moon, in the Andromeda galaxy, and any other location in the universe. The world at different moments of time can be compared to the frames of a movie, or pages of a book, following each other in a well-defined global order, the same order for the whole universe. Every object in the world "belongs" to the same "now".

In this view, if two events happen at different places at the same time (*now*), they are simultaneous for all observers. If one event at one place precedes another event at another place, their order is thought to be *absolute*—the same for all observers.

All events happening *now* represent the universe at *present*. The events of the present moment form a border between the past (all events that happened earlier) and the future (all events that will happen later). Figure 6.13 illustrates this view.

This view on the past, future, and present is *incompatible with the special theory of relativity*. Firstly, *now* is not absolute since the simultaneity of distant events is relative. Secondly, not every pair of events has an absolutely fixed *before-after* order—the events separated by a space-like interval can have any order.

According to the special theory of relativity, the separation of spacetime into past, future, and present *is local*: every pair of events must be considered separately. The spacetime around any event E_0 can be divided into three parts, depending on

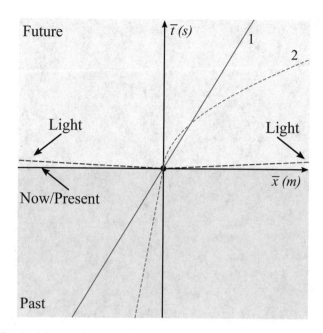

Fig. 6.13 The view of the world of events in the mechanics of Galileo and Newton. This spacetime diagram uses seconds as the units of time and meters as the units for distance, therefore the world-line of light is almost horizontal. The events making up the "world now" form a line parallel to the $\bar{x}$ axis. In three dimensional spacetime $(\bar{t}, \bar{x}, \bar{y})$, "now" would be parallel to $(\bar{x}, \bar{y})$ plane, while in four dimensional spacetime $(\bar{t}, \bar{x}, \bar{y}, \bar{z})$, "now" fills the three-dimensional space $(\bar{x}, \bar{y}, \bar{z})$. Worldlines of two objects are shown: (1) solid red line corresponds to an object moving with constant speed; (2) dashed blue line corresponds to an object accelerating after $\bar{t} = 0$

the type of the spacetime interval separating the event E_0 and any other event E. Figure 6.14 demonstrates such a partitioning.

The event E_1 lies above the light-lines and can be connected to E_0 with a time-like interval. The line connecting the events E_0 and E_1 could be a worldline of some observer traveling with $v < 1$ and being present at both events. For this observer the events will happen at the same place and will have a definite order in time. In other words, the event E_1 (and other events like it) happens *absolutely after* the event E_0. This is true for any observer, therefore the set of all events inside the top region between the light-lines is called the *absolute future* of E_0. Events in the absolute future of E_0 can be (but do not have to) influenced or caused by it.

Analogously, the event E_4 is connected to E_0 by a time-like interval and happens *absolutely before* E_0. All events below the bottom light-lines constitute *absolute past* of the event E_0. Events in the absolute past of E_0 can (but do not have to) influence or cause it.

The event E_2 lies below the top light-lines and can be connected to E_0 by a *space-like* interval. The temporal order of E_0 and E_2 is arbitrary: they can be simultaneous, E_0 can be measured as happening before E_2, or E_2 can be measured as happening

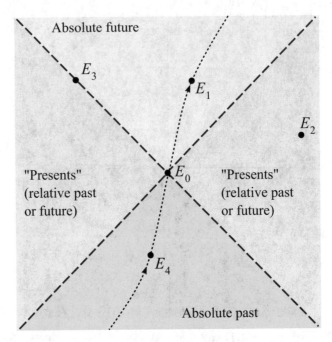

Fig. 6.14 Spacetime around any event E_0 can be split into three parts. Past is the totality of all events that may happen absolutely before the event E_0; future—all the events absolutely after E_0; present(s), or relative past or future, is comprised of all the events that may be simultaneous with the given event E_0 for some observers

before E_1. There is no *absolute* order for these events. Also, there can not be any signal, propagating with $v \leq 1$, that would connect E_0 to E_1 as cause and effect. The events like E_2 populate the part of spacetime below the top light-line and the above the bottom light-line. They have no causal connection to E_0 and represent possible "nows" (or "presents") for some observers. We may call this region "presents" or *relative past or future*, since the events from this region can be measured either as happening earlier or later than E_0, depending on the frame of reference.

Finally, all events that lie on the light-lines are clearly the events where the light signal is present. They separate absolute past or future from relative past and future for a given event. For a diagram with a single spatial dimension x, the light lines form a light-angle, splitting the spacetime region around E_0 into four visually equal parts.

For a spacetime with two spatial dimensions x and y, the Minkowski diagram requires 3 axes; light-angle becomes a *light-cone*, as shown in Fig. 6.15. In this case the separation of spacetime around any event E_0 into absolute future, absolute past, and relative past/future is similar to the two-dimensional spacetime. The events inside the top light-cone of the event E_0 form its absolute future; the events inside the bottom light-cone form the absolute past; all other events, excluding the light-cone itself, form relative past and future.

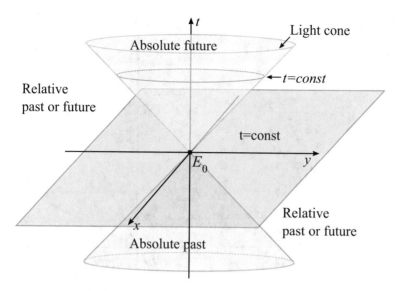

Fig. 6.15 Light cone, absolute past, absolute future, and "presents" (relative past or future) around an event E_0 in three-dimensional spacetime

Because spacetime interval is an absolute quantity, the classification of pairs of events into time-like, space-like, and light-like is independent of the frame of reference. The separation of spacetime into absolute future and absolute past around an event is also independent of the frame of reference. It is important to remember that the separation of spacetime into absolute future, absolute past and relative past/future is done *for each event individually*, in contrast to the global separation of the future from the past in pre-relativistic physics.

? Problem 6.8

Consider a particle that can travel faster than light and be present at two events E_1 and E_2 with a space-like separation. Show that there is an observer who will measure this particle as being at two places "at once".

6.7 Proper Quantities

Time interval between any two events is different in different reference frames in relative motion. Time measured relative to the given reference frame is called *coordinate time*; it is a time relative to the given spacetime coordinate system. In Minkowski diagram in Fig. 6.16a the coordinate time between the events O and

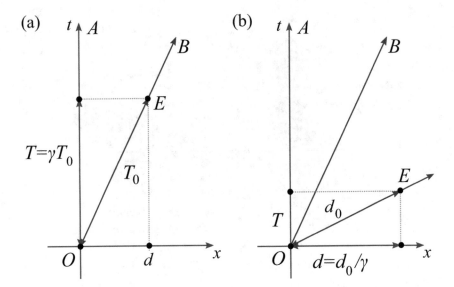

Fig. 6.16 (a) Proper time T_0 vs. coordinate time T. (b) Proper distance d_0 vs. coordinate distance d

E equals $T = t_E - t_O$. Note that these events also have *coordinate distance* $d = x_E - x_O$ between them.

The observer B, moving relative to A and present at both events O and E, will measure time between these events as T_0. For this observer the events happen in the same place—the origin of the B's reference frame. The time B measures is called *proper time*; it equals to the spacetime distance between the events O and E. The proper time and coordinate time between a pair of events are related:

$$T = \gamma T_0 = \frac{T_0}{\sqrt{1 - v^2}}, \quad T \geq T_0.$$

▷ **Proper Time**

Proper time between two events is the time measured by the clock of an inertial observer present at both events. Proper time is the spacetime distance between these events—the magnitude of the time-like spacetime interval.

Events connected with a time-like interval have a well-defined proper time. They do not have *proper distance* (see below), but they do have *coordinate distance* between them.

Proper time must not be confused with *coordinate time*. Only when a pair of events happen at the same location, the coordinate time and the proper time between these events are equal.

Distance between two events is different in different reference frames in relative motion. Distance measured relative to a given reference frame is called *coordinate distance*; it is the distance relative to a given spacetime coordinate system. In Minkowski diagram in Fig. 6.16b this is the distance $d = x_E - x_O$ measured along the x axis. The events O and E also have *coordinate time* T between them, but there is no proper time between these events!

The observer B, moving relative to A such that the events O and E are simultaneous for B, will measure the distance between these events as d_0. The distance B measures is called *proper distance* or *proper length*. The proper distance and the coordinate distance are related:

$$d = \frac{d_0}{\gamma} = d_0\sqrt{1 - v^2}, \quad d \le d_0.$$

▷ **Proper Distance**

Proper distance or *proper length* between two events is the distance measured by an inertial observer that finds these events simultaneous. Proper distance squared between these events equals to the negative spacetime interval squared:

$$\Delta s^2 = -d_0^2.$$

Events connected with space-like interval have proper distance. They do not have *proper time*, but they do have *coordinate time* between them.

Proper length represents the length of an object measured in the reference frame at rest relative to the object. *It does not represent the "true" length of an object*; see also Appendix B.5.

The terms proper time and proper length are often used in the special theory of relativity. Generally, the term *proper* means *measured at rest relative to* the object being measured: clock, rod, oscillator etc.

It is instructive to compare the proper and coordinate times to yet another time— the time between two events as "visually perceived." This is illustrated in Fig. 6.17. Light signals from two events reach the observer A as separated by time

$$T_1 = kT_0;$$

this is the familiar Doppler effect. Thus, for a pair of events, the observer A will measure time T—coordinate time; will "see" them as separated by T_1—"visual

Fig. 6.17 Proper time T_0 vs.
coordinate time T vs.
"visual" time T_1:
$T_0 \geq T \geq T_1$

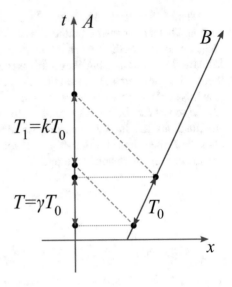

time"; whereas the proper time between the events is T_0. The following relationship
exists between these three times:[9]

$$T_0 \leq T \leq T_1.$$

Finally, for a periodic signal, such as light of a certain frequency or period, one
can speak of *proper frequency*—the frequency measured in the reference frame at
rest relative to the emitter. Proper frequency is related to the proper time between
two successive peaks in the periodic signal:

$$f_0 = \frac{1}{T_0}.$$

6.8 Lorentz Transformation as Rotation

When the axes of two Cartesian coordinate systems C and C' are rotated relative
to each other, every point gets two different sets of coordinates. As shown in the
Sect. 2.2.6, the coordinates of a point in a plane, relative to different axes, are related:

$$x' = x \cos \theta + y \sin \theta,$$

$$y' = -x \sin \theta + y \cos \theta.$$

---

[9] Recall that $k = (1 + v)\gamma$ and, therefore, $1 \leq \gamma \leq k$.

These equations have the form:

$$x' = Ax + By,$$

$$y' = Dx + Fy.$$

Where $A = F = \cos\alpha$ and $B = -D = \sin\alpha$.

Lorentz transformation can be written in a similar way:

$$t' = at + bx,$$

$$x' = dt + fx.$$

Where $a = f = \gamma$ and $b = d = -\gamma v$. The similarity with the formulas describing the rotation of Cartesian axes is evident.

There are also differences between the rotation of Cartesian axes and Lorentz transformation. Firstly, all coefficients A, B, D, and F have values from -1 to 1, whereas the coefficients a and f range from 1 to the positive infinity; the coefficients b and f range from zero to the negative infinity (for $v > 0$). Secondly, the coefficients A and B (and D and F) satisfy the relation

$$A^2 + B^2 = 1; \quad (F^2 + D^2 = 1).$$

The coefficients in Lorentz transformation satisfy the relation

$$a^2 - b^2 = 1; \quad (f^2 - d^2 = 1).$$

These relations are connected to the expressions of invariant distances in Euclidean plane and in Minkowski spacetime.

The similarities between the rotation of Cartesian axes and Lorentz transformation are deeper and more important than their differences. The similarities have an illuminating geometrical representation, as now demonstrate.

In the Sect. 6.4, we discussed the length measurement of a moving object and obtained two useful results. First, we showed that when an observer B is moving relative to an observer A, the events that happen in the same place, L_0 seconds away from the origin according to B, are described in A's reference frame with the following formula:

$$t = \frac{x}{v} + \frac{L_0}{\gamma v}.$$

As a special case, the worldline of B, corresponding to $L_0 = 0$, is described by the equation

$$t = \frac{x}{v}.$$

The events described by this equation are shown as the worldline of B in Fig. 6.18b.

Second, the events that happen at the same time T_0 according to B are described in A's reference frame with the equation

$$t = vx + \frac{T_0}{\gamma}.$$

The events simultaneous with $t' = 0$ on B's clock belong to the straight line

$$t = vx,$$

in the xt axes of the reference frame A. The events described by this equation are shown as the "coordinate axis x'" of B in Fig. 6.18b. This line has the same slope relative to x, as the line t' relative to t.

Thus, the axes t' and x' of the observer B are rotated relative to the axes of A; this rotation is symmetric with respect to the diagonal light-line. Lorentz transformation can be viewed as the rotation of spacetime axes t' and x'. The comparison between the rotation of the Cartesian axes and the Lorentz transformation is illustrated in Fig. 6.18.

? Problem 6.9

Draw Minkowski diagram from Fig. 6.18b using the reference frame of the observer B. In the diagram show all events that the observer A measures as simultaneous with $t = 0$.

6.9 "Barn and Ladder" Problem

Let us apply the results we obtained so far to the analysis of the following problem: A ladder with the proper length L_0 is lying next to a barn with the proper length $l_0 < L_0$ (See Fig. 6.19). As long as the ladder is at rest relative to the barn, or is moving with small velocity, it can not fit into the barn. However, for a large enough velocity v of the ladder, its measured length $L = L_0/\gamma$ may become smaller than l_0 and the ladder would fit fully into the barn.

When the same situation is considered in the reference frame of the ladder, the opposite conclusion must be drawn: the measured length of the barn moving with the speed v will be

$$l = l_0/\gamma < l_0 < L_0,$$

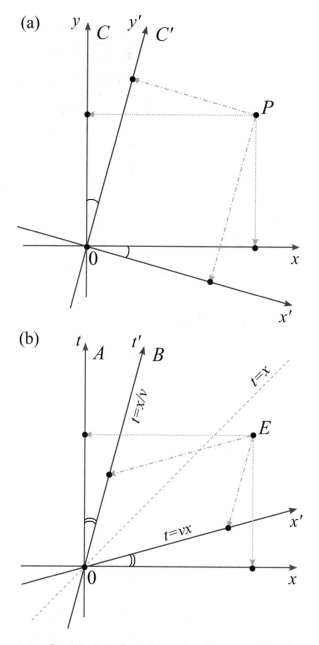

Fig. 6.18 Lorentz transformation is similar to the rotation of the axes of Cartesian coordinates: (**a**) The same point P has different coordinates when Cartesian axes are rotated relative to each other. (**b**) The same event E has different spacetime coordinates when two inertial reference frames are moving relative to each other. The Lorentz transformation looks like "rotation" of the t' and x' axes towards the worldline of light $t = x$ ($t' = x'$)

Fig. 6.19 A ladder with the proper length L_0 can not fit into the barn with the proper length l_0 if it is at rest relative to the barn. Will it fit into the barn when moving with high velocity? Will it fit from the point of view of all observers?

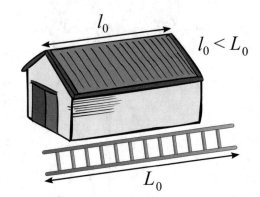

$l_0 < L_0$

and the ladder will never fully fit into the barn. Several questions arise: *Will the ladder fully fit into the barn or will it not? Why do different observers give different answers? Isn't it an absolute fact that one thing fits inside the other?*

6.9.1 Analysis in Barn's Reference Frame

Figure 6.20 shows Minkowski diagram for the ladder moving through the barn. The ladder will fit into the barn if the measured length L of the ladder is smaller than the measured length of the barn:

$$L < l_0,$$

here $L = L_0/\gamma$ is the length of the moving ladder, measured in the barn's reference frame. Substituting L into the inequality, we get the condition for the paradox

$$\frac{L_0}{l_0} < \gamma. \tag{6.21}$$

This condition determines the minimum velocity that will result in $L < l_0$ (the moving ladder fits inside the barn). When this happens, the front of the ladder will still be inside the barn when the back of the ladder enters the front door (event 3 on the diagram in Fig. 6.20).

Obviously, the front of the ladder can not exit the barn before it enters the barn, and therefore the event 2 is absolutely after 1. Similar statement can be made about the pairs of events 4 and 3, 3 and 1, 4 and 2. The intervals between these pairs of events are causally connected, as reflected by their time-like character. Indeed, there are worldlines connecting those pairs of events, corresponding to material particles moving through them (the front of the ladder for the pair 1 and 2, for example).

Fig. 6.20 Spacetime diagram of the ladder moving through the barn. The reference frame of the barn is used, with the worldlines of the front (F) and back (B) doors shown. In the barn's frame of reference the measured length of the ladder is $L = L_0/\gamma$; it can be arbitrarily small for a suitable choice of the speed $v < 1$

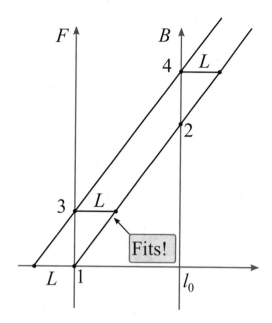

In contrast, the events 2 and 3 must be connected by the space-like interval, if we require the ladder to fully fit inside the bard—it follows from the requirement for the paradox, as will be demonstrated shortly. It means that the events 2 and 3 have no absolute relationship and their order is relative: in some reference frames 2 happens before 3, in others—3 before 2, and in one particular reference frame they happen at the same time. To see this, we first need the spacetime coordinates of both events.

From the Minkowski diagram we find

$$(2) \quad \rightarrow \quad \left(\frac{l_0}{v}, l_0\right),$$

$$(3) \quad \rightarrow \quad \left(\frac{L}{v}, 0\right) = \left(\frac{L_0}{v\gamma}, 0\right).$$

The distance squared between these events is

$$(\Delta x)^2 = l_0^2;$$

the time squared

$$(\Delta t)^2 = \left(\frac{l_0}{v} - \frac{L_0}{v\gamma}\right)^2 = \frac{l_0^2}{\gamma^2 v^2}\left(\gamma - \frac{L_0}{l_0}\right)^2.$$

It can easily be shown that

$$\gamma^2 v^2 = \gamma^2 - 1 = (\gamma - 1)(\gamma + 1),$$

and, using the fact that $l_0^2 = (\Delta x)^2$, the time squared is written as:

$$(\Delta t)^2 = (\Delta x)^2 \left[\frac{\gamma - L_0/l_0}{\gamma - 1} \right] \times \left[\frac{\gamma - L_0/l_0}{\gamma + 1} \right].$$

The fraction in the first brackets is positive and less than unity; indeed, since $L_0 > l_0$, the number greater than 1, but less than γ—as required by the condition of the paradox (6.21)—is subtracted from γ on top. The fraction in the second brackets is less than unity (but also greater than zero), because we subtract from the numerator, while adding to the denominator. We conclude that

$$(\Delta t)^2 < (\Delta x)^2 \quad \rightarrow \quad (\Delta t)^2 - (\Delta x)^2 < 0 \qquad (6.22)$$

and the interval between the events 2 and 3 is *space-like*. These events can not be causally connected and can have different order for different observers. It is important to remember that the separation between the events 2 and 3 *is not always space-like*; it becomes space-like when the ladder moves fast enough to fit into the barn.

In summary, the ladder will fit for some observers and will not fit for others. The expression *"the ladder fits"* is a statement about its length and relies on the positions of two distant points *at the same time*. It is not a statement about a single event, but about two events that are not absolutely related. The common sense expectation that the order of *any* two events must be the same for all observers is wrong, it causes the feeling of a paradox in this particular problem and similar problems that rely on simultaneity of distant events. A true paradox would happen if, for example, the back of the ladder left the barn before the front of the ladder.

Chapter Highlights

- The fundamental invariant in the special theory of relativity is given by the equation $s^2 = t^2 - x^2$. It serves as the distance in spacetime. Three kinds of intervals are possible: time-like, space-like, and light-like.
- Not every pair of events has an absolute order in time. Only events connected by a time-like interval have absolute order.
- The global separation of spacetime into "past", "present", and "future" is not absolute. Such separation can be done around each event individually.
- Length of an object, as well as the concept of an "object", relies on the idea of simultaneity and thus has no absolute character in the special theory of relativity. The length of a moving object is less than the length at rest: $L = L_0\sqrt{1 - v^2}$. Proper length *does not represent* the "true" length.

References

1. Minkowski, H. (1952). Space and time. In *The Principle of Relativity*. New York: Dover.
2. Born, M. (1965). *Einstein's Theory of Relativity* (pp. 253–255). New York: Dover
3. FitzGerald, G. F. (1889). The Ether and the Earth's Atmosphere. *Science 17, 13*(328), 390.
4. Lorentz, H. A. (1892). *The Relative Motion of the Earth and the Aether*. Amsterdam: Zittingsverslag Akad. v. Wet.
5. Viereck, G. S. (1929). *What Life Means to Einstein, the Saturday Evening Post*.

Chapter 7
Special Relativity C

> *Measurements of distances and times do not directly reveal properties of the things measured, but relations of the things to the measurer. What observation can tell us about the physical world is therefore more abstract than we have hitherto believed.*
>
> B. Russell, *ABC of Relativity.*

Abstract This chapter covers more advanced topics of the special theory of relativity: motion in two dimensions, spacetime vectors, momentum and energy, and accelerated motion.

7.1 Lorentz Transformation in 4D

We derived Lorentz transformation connecting spacetime coordinates (t, x) and (t', x') of the same event, as measured by different observers in relative motion:

$$t' = \gamma(t - vx)$$
$$x' = \gamma(x - vt).$$

It is important to remember that these equations have been obtained under the following assumptions:

1. The relative motion is happening along the direction of the axes x and x'.
2. At the moment $t = t' = 0$ the origins of the axes x and x' of the observers coincide: $x = x' = 0$.

Nothing has been said about the measurements of the remaining coordinates (y, z or y' z') of events. We will find the relationship between the remaining coordinates next.

© The Author(s), under exclusive license to Springer Nature Switzerland AG 2022
Y. Deshko, *Special Relativity*, Undergraduate Lecture Notes in Physics,
https://doi.org/10.1007/978-3-030-91142-3_7

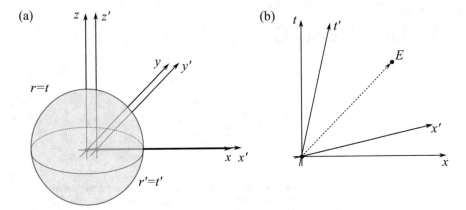

Fig. 7.1 (a) A flash of light happens at the origin of two reference frames in relative motion. (b) Minkowski diagram of a light pulse propagating in one direction along the x axis. Spacetime axes for two observers in relative motion as drawn as (t, x) and (t', x')

Let us first consider a simple case when a flash of light happens at $t = 0$ and $x = 0$. Light will propagate in all directions uniformly, with the same speed for both observers. The situation is illustrated in Fig. 7.1a. The distance from the origin covered by light in any direction is given by

$$r = t, \quad r' = t',$$

for the first and the second observer, respectively. From these relations follows that

$$x^2 + y^2 + z^2 = t^2, \quad x'^2 + y'^2 + z'^2 = t'^2.$$

Using the fact that $t^2 - x^2 = t'^2 - x'^2$ is an invariant, we obtain

$$y^2 + z^2 = y'^2 + z'^2.$$

If we allow a possibility of $y' < y$, then the requirement $z' > z$ follows. But why would the measurement of y' be affected in a way different from z'? After all, the orientation of the axis y' is arbitrary. Similar issue arises if we allow $y' > y$ (then it must be $z' < z$). The only possibility, not leading to a contradiction, is that

$$y' = y$$
$$z' = z.$$

These relations are true for all events connected to the event $(0, 0)$ via a light-like interval, as shown in Fig. 7.1b.

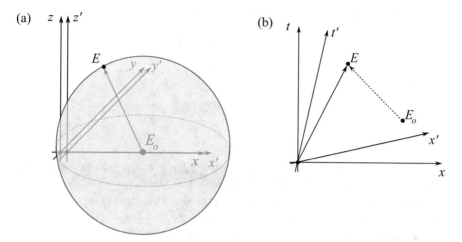

Fig. 7.2 (**a**) A flash of light happens at an event E_0 on the axis x and reaches an arbitrary event E. (**b**) Minkowski diagram of the light propagating from the event E_0 to E. The axes of two reference frames in relative motion are also shown

Next, consider a more general case of an event E not directly connected by a light-like interval to the origin. The case is illustrated in Fig. 7.2. An event E has spacetime coordinates (t, x, y, z) in the reference frame S, and (t', x', y', z') in the reference frame S'.

For any event we can draw a past light cone, consisting of all events which could send light to E. Let us choose one such event on the past light cone, indicated in Fig. 7.2b as E_0. The only requirement we place on this event is that it lies on the x axis;[1] see Fig. 7.2a. The spacetime coordinates of the events E and E_0 are

$$E \rightarrow (t, x, y, z) \text{ or } (t', x', y', z')$$
$$E_0 \rightarrow (t_0, x_0, 0, 0) \text{ or } (t'_0, x'_0, 0, 0).$$

Since the events E_0 and E are connected via a light-like interval, we can write

$$(t - t_0)^2 = (x - x_0)^2 + y^2 + z^2$$

for the observer in S, and

$$(t' - t'_0)^2 = (x' - x'_0)^2 + y'^2 + z'^2$$

for the observer in S'.

---

[1] It should be clear that if the event E_0 lies on the x axis, it must lie on the x' axis as well.

Using the invariance of the interval

$$(t - t_0)^2 - (x - x_0)^2 = (t' - t_0')^2 - (x' - x_0')^2$$

once more, we get

$$y^2 + z^2 = y'^2 + z'^2.$$

We make the following conclusion:

▷ **Lorentz transformation in 4D**

The measurements of coordinates and distances along the directions perpendicular to the relative motion are not affected by the motion.

The Lorentz transformation for all coordinates is given by

$$t' = \gamma(t - vx)$$
$$x' = \gamma(x - vt)$$
$$y' = y$$
$$z' = z.$$

A less formal demonstration that the measurements in the directions perpendicular to the direction of the relative motion are not affected, can be given using a situation depicted in Fig. 7.3. A ring with the inner radius R is allowed to slide along a rod with the same radius R. If we allow the measurements of, say, coordinate y to be affected, then the relative motion will be possible in one reference frame and impossible in the other.

To see this, assume that due to the relative motion the distances in y direction "shrink". Then, in the reference frame S of the ring, the rod is measured to have an ellipsoid shape, with the y dimension smaller than $2R$. The rod will freely move relative to the ring. In the reference frame S' of the rod, however, the ring's size in the y direction is affected, making the inner hole of the ring narrower than $2R$. The ring can not move along the rod.

Similar arguments can be used to conclude that transverse distances in any directions must be unaffected.

Fig. 7.3 Motion of a ring
along a rod must be possible
from the standpoint of any
observer. This requires that
the measurements of
coordinates and distances in
the directions perpendicular
to the relative motion are not
affected by it

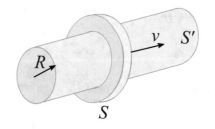

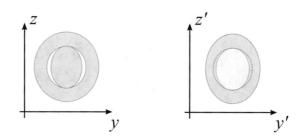

Find the Lorentz transformation expressing

$$(t, x, y, z)$$

in terms of

$$(t', x', y', z').$$

7.2 Velocity Composition Revisited

Full Lorentz transformation of spacetime coordinates (t, x, y, z) show that the
coordinates along the axes perpendicular to the direction of relative motion are not
affected. However, we should not get an impression that nothing related to these
directions is affected.

Consider a particle moving relative to the reference frame S'. During a time
interval $\Delta t'$ its position along each axis chages by

$$\Delta x', \ \Delta y', \ \Delta z',$$

and corresponding components of the velocity, relative to S', are

$$u'_x = \frac{\Delta x'}{\Delta t'}, \; u'_y = \frac{\Delta y'}{\Delta t'}, \; u'_z = \frac{\Delta z'}{\Delta t'}.$$

The motion of the same particle, when observed in the reference frame S, results in the change of its coordinates along each axis by

$$\Delta x, \; \Delta y, \; \Delta z,$$

during a time interval $\Delta t \neq \Delta t'$. Using the Lorentz transformation

$$\Delta t = \gamma(\Delta t' + v\Delta x'),$$

$$\Delta x = \gamma(\Delta x' + v\Delta t'),$$

$$\Delta y = \Delta y',$$

$$\Delta z = \Delta z',$$

we can find all velocity components as measured by the observer S:

$$u_x = \frac{\Delta x}{\Delta t} = \frac{\Delta x' + v\Delta t'}{\Delta t' + v\Delta x'} = \frac{u'_x + v}{1 + vu'_x},$$

$$u_y = \frac{\Delta y}{\Delta t} = \frac{\Delta y'}{\gamma(\Delta t' + v\Delta x')} = \frac{u'_y}{\gamma(1 + vu'_x)},$$

and

$$u_z = \frac{\Delta z}{\Delta t} = \frac{\Delta z'}{\gamma(\Delta t' + v\Delta x')} = \frac{u'_z}{\gamma(1 + vu'_x)}.$$

From these equations, it is clear that, in general, $u_y \neq u'_y$ and $u_z \neq u'_z$.

Example:
Consider a particle moving along the y' axis with the velocity $u'_y = U$ in the reference frame S'. In the reference frame S, the particle will be moving along the x axis with the velocity

$$u_x = \frac{u'_x + v}{1 + vu'_x} = v,$$

(continued)

and along the y axis with the velocity

$$u_y = \frac{u'_y}{\gamma(1 + vu'_x)} = \frac{U}{\gamma}.$$

Compared to the predictions of Galileo-Newtonian mechanics, the motion along the x axis is the same, while the motion along the y axis is *slower*.

? Problem 7.2

Express the components of velocity u'_x, u'_y, u'_z in terms of u_x, u_y, u_z.

? Problem 7.3 *

Consider a case when $u'_z = 0$, $u'_x = \cos\alpha$, and $u'_y = \sin\alpha$. This corresponds to a photon traveling in the $(x'y')$-plane at an angle α relative to the axis x'. Check that the magnitude of the speed of the photon, relative to S, is unity. That is, check that $u_x^2 + u_y^2 = 1$.

7.3 Relativistic Aberration

The formulas for the velocity composition can be used to analyze an interesting and important phenomenon of light propagation, observed in different reference frames. The situation is illustrated in Fig. 7.4, where a ray of light is moving along a straight line making an angle θ with the horizontal axis x in the reference frame S.

If we consider the ray of light as a collection of particles (photons), then each photon will have the following velocity components:

$$u_x = -c\cos\theta = -\cos\theta,$$

$$u_y = -c\sin\theta = -\sin\theta.$$

In the reference frame S', moving relative to S with the velocity v, the components will be given by the formulas

$$u'_x = \frac{u_x - v}{1 - vu_x},$$

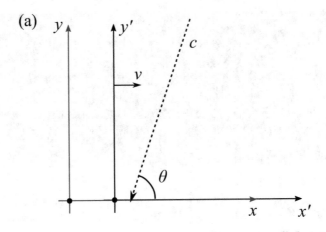

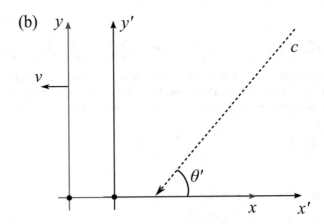

Fig. 7.4 (**a**) A ray of light traveling from a distant star observed in a reference frame S. The ray is traveling along a straight line, having an angle θ with the axis x. (**b**) The same ray of light observed in a reference frame S' moving relative to S. The ray of light still travels along a straight line but with a different slope

$$u'_y = \frac{u_y}{\gamma(1 - vu_x)}.$$

Substituting the expressions for u_x and u_y, we get

$$u'_x = -\frac{v + \cos\theta}{1 + v\cos\theta},$$

$$u'_y = -\frac{\sin\theta}{\gamma(1 + v\cos\theta)}.$$

Since both u'_x and u'_y are constant, the trajectory of the light, as observed in the reference frame S', will be a straight line, shown in Fig. 7.4b. The speed of light being the same in all frames of reference, we can write

$$u'_x = -\cos\theta', \quad u'_y = -\sin\theta'.$$

The relation between the angles of the same ray of light, as observed in the reference frames S and S', can be expressed as follows:

$$\cos\theta' = \frac{v + \cos\theta}{1 + v\cos\theta},$$

$$\sin\theta' = \frac{\sin\theta}{\gamma(1 + v\cos\theta)}.$$

It is helpful to study some special cases.

Example:
Consider a pair of observers, S and S', moving with the relative velocity $v = 0.9$. In the reference frame of the observer S there are four rays of light, having the slope angles of $\theta_1 = 10°$, $\theta_2 = 45°$, $\theta_3 = 90°$, and $\theta_4 = 120°$. In the reference frame S' the slope angles will be $\theta'_1 = 2.30°$, $\theta'_2 = 10.9°$, $\theta'_3 = 25.8°$, and $\theta'_4 = 43.3°$.

Figure 7.5 shows how the light, propagating uniformly from all directions in the reference frame S, becomes more and more confined to a small angle in the reference frame S'.

As demonstrated in Fig. 7.5, for a rapidly moving observer the incoming light will be confined to a small range of angles around $\theta' = 0$. All light will be coming mostly along the axis x'. In addition, the frequency of the received light will be higher than the frequency of the emitted light (optical Doppler effect discussed earlier). A crew in a spaceship moving with $v = 0.99$ will not see a starry sky with yellowish dots uniformly scattered in every direction. Instead, an extreme ultraviolet radiation will be incoming in a narrow cone of about $10°$ around the direction of motion relative to the distant stars.

Figure 7.6 shows how the angle, measured in the reference frame S' (which is moving relative to S with speed v), depends on the relative velocity v. The rays considered have angles $90°$, $60°$, $45°$, $30°$ and $5°$, in the reference frame S.

? Problem 7.4

Consider a ray of light coming vertically in the reference frame of S ($\theta = 90°$). Assume that the relative velocity of the reference frame S' is small: $v \ll 1$ and

$\gamma \approx 1$. Show that in this case, in the reference frame of S', the light is coming at an angle θ' such that

$$\tan \theta' = \frac{c}{v}.$$

This is the same expression that one would obtain using Galilean velocity addition.

Fig. 7.5 Relativistic aberration effect for three different relative velocities. In the reference frame S, light rays are coming from all directions, with the sources uniformly distributed. In the reference frame S', the distribution of the directions of light rays changes. Increasing the relative velocity leads to light coming predominantly from a small angle around the axis x'

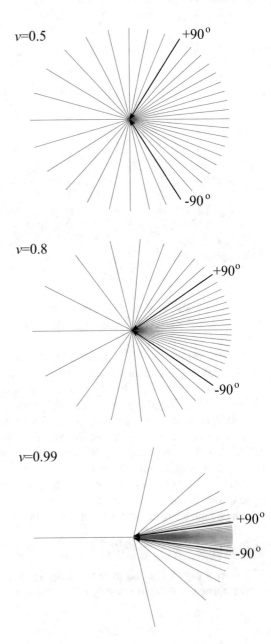

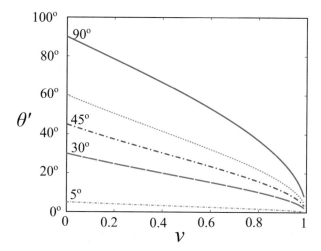

Fig. 7.6 Relativistic aberration effect, observed in the reference frame S', for several directions of light propagation in S, as the function of relative velocity of two observers. The rays of light, coming from the directions of 90°, 60°, 45°, 30°, and 5° in the reference frame S, have different directions in S'; the rays are getting closer to the axis x', as the relative velocity v of S and S' increases

The effect of starlight aberration due to the motion of the Earth relative to the distant stars was discovered by an English astronomer James Bradley in 1727.

7.3.1 Relativistic Beaming

An effect closely related to aberration happens when a *source of electromagnetic radiation is moving* with high speed relative to an observer.

Consider a source of light, emitting uniformly in all directions, as shown in Fig. 7.7a. When the same source is moving with the speed $v \approx 1$, the distribution of the radiation becomes very different. To find it, we can use the same arguments that were used to describe the relativistic aberration. The conclusion, illustrated in Fig. 7.7b for $v = 0.99$, is that the source will emit electromagnetic radiation predominantly in the direction of its motion—in the manner of a "searchlight". This is the effect of *relativistic beaming*, also called "headlight" or "searchlight" effect.

The effect of beaming becomes important whenever a source of electromagnetic radiation is moving with speed close to the speed of light. Examples include charged particles (electrons, protons) in accelerators, or astrophysical phenomena involving plasma rapidly falling into compact massive astronomical objects.

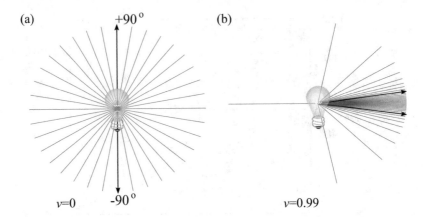

Fig. 7.7 Relativistic beaming: radiation pattern changes when the source is moving with high speed. Instead of emitting uniformly in all directions, as in (**a**), the source will emit predominantly in the direction of its motion, as in (**b**)

7.4 "Momentum" Not Conserved

The transformation equations for the velocity components u_x, u_y, and u_z lead to the following important result:

▷ **Newtonian Momentum Not Conserved**

The physical quantity, given by the equation

$$p = mv,$$

is *not conserved* in collisions. The expression for momentum in Newtonian mechanics does not agree with high-energy collision experiments, nor with the special theory of relativity.

To see this, we can analyze the problem of symmetric collision, considered from the point of view of Newtonian mechanics in the Appendix B.1. For convenience, the situation is illustrated again in Fig. 7.8. Two identical observers in relative motion carry identical particles. Each observer throws a particle perpendicular to the relative motion. The particles collide elastically and bounce back into the hands of the observers.

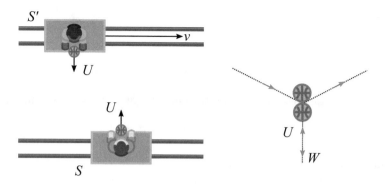

Fig. 7.8 Symmetric elastic collision of two "particles" in case of large relative speed v of two observers. Using the Newtonian expression for momentum, $p = mv$, results in the violation of the momentum conservation in this collision

The key point is that the observers are identical and both use identical particles. The observer S has a particle (1), moving with the velocity components

$$u_{1x} = 0, \ u_{1y} = U$$

before the collision, and

$$u_{1x} = 0, \ u_{1y} = -W$$

after. Similarly, the observer S' has a particle (2), moving with the velocity components (relative to the $x'y'$ axes):

$$u'_{2x} = 0, \ u'_{2y} = -U$$

before the collision, and

$$u'_{2x} = 0, \ u'_{2y} = W$$

after.

Since both reference frames are equivalent, we can use the reference frame S to check the conservation of "momentum" $p = mv$ during the collision. Given the velocity components of the particle (2) relative to the S', we can find the components relative to S, using the relativistic formulas for the velocity component transformation. Before the collision:

$$u_{2x} = v,$$

$$u_{2y} = -U/\gamma,$$

and after:

$$u_{2x} = v,$$

$$u_{2y} = W/\gamma.$$

The conservation of Newtonian momentum along the y axis looks as follows:

$$mU - mU/\gamma = -mW + mW/\gamma.$$

Simplifying this expression results in

$$mU\frac{\gamma - 1}{\gamma} = mW\frac{-\gamma + 1}{\gamma}.$$

The left-hand side is positive, while the right-hand side is negative; therefore, the conservation of "momentum" $p = mv$ is not satisfied.

As shown in a more detailed analysis in the Appendix B.2, the correct relativistic expression for the momentum of a particle is

$$P(v) = \gamma mv = \frac{mv}{\sqrt{1 - v^2}}.$$

It is tempting to write this expressions in a Newtonian form

$$P(v) = M(v)v,$$

where

$$M(v) = \frac{m}{\sqrt{1 - v^2}},$$

and interpret the last equation as if the mass of the object changes with speed. *We should not make this mistake, because in the special and general theory of relativity mass of a particle is an invariant quantity.*

In a letter to an editor and author Lincoln Barnett, Einstein wrote:[2]

▷ **Einstein on Mass**

It is not good to introduce the concept of the mass $M = m/\sqrt{1 - \bar{v}^2/\bar{c}^2}$ of a moving body for which no clear definition can be given. It is better to introduce no other mass concept that the "rest mass" m. Instead of introducing M it is better to mention the expression for the momentum and energy of a body in motion.

---

[2] Translation quoted from *The Concept of Mass*, Lev B. Okun, Physics Today, June 1989.

The concept of mass has rich history, the traces of which can still be found in many books about special relativity. One trace is the idea that "mass of a moving object increases with its speed". Taylor and Wheeler in their book *"Spacetime Physics"* call this *"abuse of the concept of mass"*.[3] We avoid such an abuse in this book.

Why did people think that mass may be velocity dependent? One reason is the use of wrong formula for the momentum of an object. Suppose we perform an experiment where we apply a known force F to a particle during a fixed short interval of time δt (see Appendix A.2 for the use of δ-notation). According to Newton's second law, we will change the momentum of the particle by an amount proportional to $F\delta t$. Indeed, this is the second law of mechanics:

$$F = \frac{\delta p}{\delta t} = \frac{\delta(mv)}{\delta t} = m\frac{\delta v}{\delta t} = ma.$$

Using the Newtonian formula for momentum $p = mv$, we can measure the mass of a particle if we measure the change of velocity due to a known force pulse $F\delta t$:

$$m = \frac{F\delta t}{\delta v}.$$

The momentum of the particle, however, is given by the relativistic formula

$$P = \gamma mv. \quad (P \geq mv)$$

Let us consider, as an illustration, the case of intermediate velocity, when the approximation

$$\gamma \approx 1 + \frac{v^2}{2}$$

can be used, so that the momentum is given by

$$P \approx mv + \frac{mv^3}{2}.$$

To change the relativistic momentum by the same magnitude $F\delta t$, a smaller change in velocity δv_{rel} is required, compared to the Newtonian case. Therefore, in the measurement with a given force pulse $F\delta t$, the observed change of velocity will be smaller than expected from the Newtonian mechanics:

$$\delta v_{rel} < \delta v_{Newt},$$

---

[3] E. Taylor, A. Wheeler *Spacetime Physics*, Freeman and Company, 2nd ed., p. 246.

and the inferred "mass" will be

$$m_{rel} = \frac{F\delta t}{\delta v_{rel}} > \frac{F\delta t}{\delta v_{Newt}} \quad (= m).$$

Moreover, the result, m_{rel}, will depend on the velocity v.

Experiments along these lines were performed by the German physicists Kaufmann and Bucherer shortly after the discovery of fast moving electrons. They studied how electrons, emitted by a radioactive material, were deflected in magnetic fields. Kaufmann and Bucherer showed that the "mass" p/v depends on velocity.

The electron discovery raised fundamental questions: *What is the nature of electron? Where does its mass come from?* The theory of electromagnetism provided some answers, which dependent on a particular model for electron. Without going into details, we just mention that several models existed, predicting different formulas for the mass of an electron as the function of its velocity. The experiments performed by Kaufmann, Bucherer, and others aimed at establishing which model worked best.

Today we know that Newtonian expression for the momentum and kinetic energy are only approximately correct. We also know that mass of any object, including electron, is a relativistic *invariant* and it is *independent of the object's velocity*.

7.5   Spacetime Vectors

A point in a Euclidean plane can be specified by a pair of coordinates, for example Cartesian (x, y) or polar (r, ϕ). Alternatively, the same point can be specified by an *arrow* connecting the origin of the coordinate system and the point in question. Both descriptions are equivalent, both provide two different representation of the same object—a *vector*. There are many vector quantities used in physics: position, velocity, acceleration, force, electric and magnetic field, current density, to name a few.

It is important to understand the following fact about vectors: They are not reduced to a set of numbers, or to arrows; the former and the latter are merely useful *representations* of an abstract mathematical concept that has many "faces".

An event in spacetime can be specified similarly to a point in a plane: using a pair of coordinates (t, x), measured in some reference frame, or an arrow connecting the origin O of a reference frame and the event of interest. The "length" of such an arrow is given by the interval between the given event and the event O. The similarity and difference between the geometries in a plane and two-dimensional spacetime are highlighted in Fig. 7.9.

Lorentz transformation specifies how the coordinates (t, x) of the same event transform when we change reference frame. The transformation of spacetime coordinates of an event is mathematically similar to the transformation of the Cartesian coordinates (x, y) of a point in a plane, when the coordinate axes

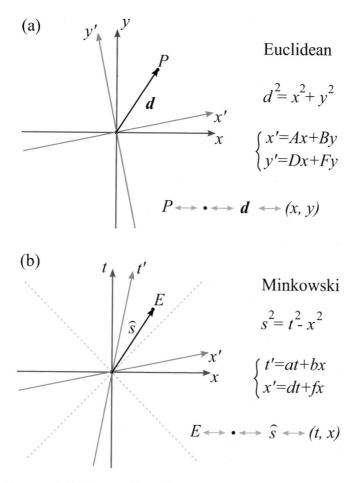

Fig. 7.9 There are similarities, as well as differences, between Euclidean plane and Minkowski spacetime. The geometry of spacetime is not Euclidean. This is reflected in the expression for the invariant "distance", and also in the different type of coordinate transformation—Lorentz transformation. In spacetime we can introduce vectors that have coordinates (t, x) as components. Such vectors are analogous to position vectors in Euclidean plane

are rotated. The spacetime coordinates (t, x) can be viewed as representing the components of a *spacetime vector*, which represents the position of some event.

To avoid the clash of notations, we will denote spacetime vectors differently from "regular" vectors. *Spacetime position* will be denoted like this:

$$\widehat{s} = (t, x) = (s_t, s_x),$$

where s_t is the *temporal component* of the spacetime position vector $\widehat{s}$, and s_x is its *spatial component*.

The notation for spacetime vectors with an arc over a letter is *not* standard. However, it does not introduce any difficulties and helps distinguish "regular" vectors from vectors in spacetime.

7.5.1   Spacetime Displacement

For two events specified by spacetime positions

$$\overset{\frown}{s}_1 = (t_1, x_1) \text{ and } \overset{\frown}{s}_2 = (t_2, x_2)$$

we calculate the *spacetime displacement*

$$\Delta\overset{\frown}{s} = \overset{\frown}{s}_2 - \overset{\frown}{s}_1$$

as follows:

$$\Delta\overset{\frown}{s} = (\Delta s_t, \Delta s_x) = (\Delta t, \Delta x) = (t_2 - t_1, x_2 - x_1).$$

In other words, the difference of spacetime positions is calculated component-wise, similar to the addition and subtraction of "regular" vectors.

This procedure is illustrated in Fig. 7.10. The spacetime displacement $\Delta\overset{\frown}{s}$ is just as good a vector as $\overset{\frown}{s}_1$ and $\overset{\frown}{s}_2$. Similar to $\overset{\frown}{s}_1$ and $\overset{\frown}{s}_2$, it is represented by an arrow in Minkowski plane. Furthermore, its components, Δs_t and Δs_x, are transformed between different refence frames in the same way as the components of $\overset{\frown}{s}_1$ and $\overset{\frown}{s}_2$:

$$\Delta s'_t = \gamma(\Delta s_t - v\Delta s_x),$$
$$\Delta s'_x = \gamma(\Delta s_x - v\Delta s_t).$$

These relationships follow directly from the definitions of Δs_t and Δs_x, and the transformation formulas for t and x. The important conclusion is this: *The components of the spacetime displacement vector $\Delta\overset{\frown}{s}$ in different reference frames are connected via Lorentz transformation.*

Depending on the pair of events it connects, a displacement vector can correspond to a time-like, space-like, or a light-like interval. The invariant quantity Δs, defined as

$$\Delta s^2 = \Delta t^2 - \Delta x^2 = \Delta s_t^2 - \Delta s_x^2, \tag{7.1}$$

is a natural fit for the notion of the *length of the spacetime displacement vector*.

(a)   (b)

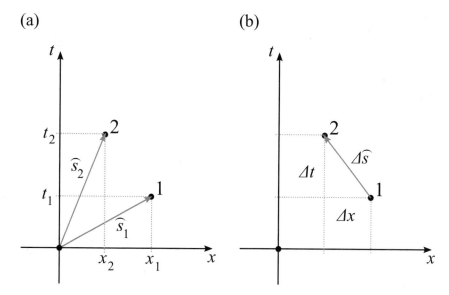

Fig. 7.10 Given two events, specified by their spacetime positions $\widehat{s}_1$ and $\widehat{s}_2$, we can calculate the spacetime displacement $\Delta\widehat{s}$

7.5.2 Spacetime Velocity

The Galileo-Newtonian velocity

$$\mathbf{v} = \frac{\Delta \mathbf{d}}{\Delta t} = \left(\frac{\Delta x}{\Delta t}, \frac{\Delta y}{\Delta t} \right)$$

has the spacetime counterpart, constructed using the spacetime displacement and proper time:

$$\widehat{u} = \frac{\Delta \widehat{s}}{\Delta T_0} = \left(\frac{\Delta t}{\Delta T_0}, \frac{\Delta x}{\Delta T_0} \right).$$

The spacetime velocity is defined *only for time-like displacements*, when $\Delta\widehat{s}$ represents a piece of a worldline of a moving object, such as the one in Fig. 7.11. In this case the proper time ΔT_0 equals to the length Δs of the vector $\Delta\widehat{s}$.

In terms of spacetime components, the spacetime velocity can be written as

$$\widehat{u} = (\frac{\Delta t}{\Delta T_0}, \frac{\Delta x}{\Delta T_0}) = (\frac{\Delta t}{\Delta s}, \frac{\Delta x}{\Delta s}).$$

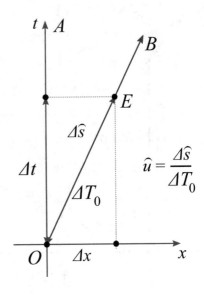

Fig. 7.11 Spacetime velocity equals to the spacetime displacement per unit of proper time. Coordinate time can not be used in the definition of the spacetime velocity, because it is not an invariant quantity

Recalling the connection between the coordinate time and proper time

$$\Delta T_0 = \Delta s = \frac{\Delta t}{\gamma},$$

we obtain

$$\widehat{u} = (\gamma, \gamma u) = (u_t, u_x),$$

where

$$u = \frac{\Delta x}{\Delta t}$$

is the usual velocity of Newtonian mechanics, and

$$\gamma = \gamma(u) = \frac{1}{\sqrt{1 - u^2}}.$$

An interesting interpretation of the relativistic factor γ follows from the components of the spacetime velocity: γ is the time component of spacetime velocity of an object. In more colorful, but less precise, language—γ tells how fast an object is moving "through time".

From the definition of spacetime velocity $\widehat{u}$

$$\widehat{u} = \frac{\Delta \widehat{s}}{\Delta T_0} = \frac{1}{\Delta T_0} \Delta \widehat{s}$$

it is evident that $\widehat{u}$ is obtained from $\Delta\widehat{s}$ by multiplying it by a number $1/\Delta T_0$. Importantly, this number is *an invariant*—the same for all reference frames; therefore, the quantity $\widehat{u}$ has the same vector character as $\Delta\widehat{s}$. The velocity $\widehat{u}$ points in the same direction as the displacement $\Delta\widehat{s}$, but its length is scaled by a factor $1/\Delta T_0$.

The length squared of spacetime velocity is proportional to the length squared of the spacetime displacement:

$$u^2 = \frac{1}{\Delta T_0^2}\Delta s^2.$$

Using the fact that $\Delta T_0 = \Delta s$, we conclude that

$$u^2 = 1.$$

Note: We must not confuse the u^2 here with the square of the coordinate velocity $u = \Delta x/\Delta t$. Sometimes, to avoid mixing up the two, the magnitude squared of a vector $\widehat{u}$ is written as

$$|\widehat{u}|^2.$$

This notation is not used here, since we do not talk about vector magnitudes too much.

In terms of spacetime components

$$u_t = \frac{\Delta t}{\Delta T_0}, \; u_x = \frac{\Delta x}{\Delta T_0},$$

the length squared of $\widehat{u}$ can be written as follows:

$$u^2 = \frac{\Delta t^2 - \Delta x^2}{\Delta T_0^2} = u_t^2 - u_x^2.$$

What is the use of a vector that always has the same length? A vector with the same length can have different directions, as illustrated in Fig. 7.12. The change of direction signifies an acceleration due to an interaction; the faster the worldline "bends", the more "curvy" it is—the stronger the interaction.

Note that the spacetime velocity is obtained by dividing a spacetime displacement vector $\Delta\widehat{s}$ by a number—the magnitude of the interval—an invariant quantity. For comparison, if we write a quantity

$$\frac{\Delta\widehat{s}}{\Delta t}$$

Fig. 7.12 Spacetime velocities of objects, moving with different speeds relative to the observer A, have different directions. The directions of the spacetime velocities of the objects B and C, moving uniformly and rectilinearly, remain unchanging. The direction of the spacetime velocity of an accelerating object D is gradually changing. All spacetime velocities have the same magnitude 1

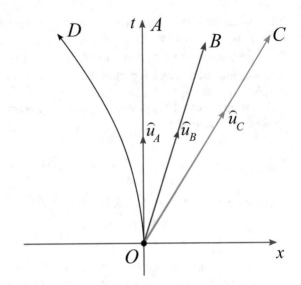

it *will not be a spacetime vector*, at least because its magnitude will be different in different reference frames. Indeed, for a given pair of events the magnitude of spacetime displacement $\Delta \widehat{s}$ is an invariant quantity (interval), whereas the coordinate time Δt between those events is different for different reference frames. The situation is analogous to the Euclidean plane, where not every pair of numbers represents a vector. The displacement $\Delta \mathbf{d} = (\Delta x, \Delta y)$ is a well-behaved vector, while the quantity

$$\xi = \frac{\Delta \mathbf{d}}{\Delta x} = (1, \frac{\Delta y}{\Delta x}),$$

can not correspond to a vector: its length depends on the choice of a coordinate system. For some coordinate system, the axis y can go through the pair of points connected by $\Delta \mathbf{d}$; in this case $\Delta y = 0$ and the length squared of ξ will be

$$\xi^2 = 1.$$

For all other orientations of the axes, $\Delta y > 0$ and the length squared becomes

$$\xi^2 = 1 + \left(\frac{\Delta y}{\Delta x}\right)^2 > 1.$$

In general, to specify an event we require four spacetime coordinates (t, x, y, z). Correspondingly, spacetime displacement and spacetime velocity will have four components. For example:

$$\widehat{u} = (u_t, u_x, u_y, u_z).$$

The quantity $\widehat{u}$ is often called *four-velocity*, due to the number of components.

7.5.3 Lorentz Transformation of Velocity

The spacetime coordinates of the same event in different reference frames are related via Lorentz transformation:

$$t' = \gamma(t - vx),$$
$$x' = \gamma(x - vt).$$

The same transformation connects the *change in spacetime coordinates*:

$$\Delta t' = \gamma(\Delta t - v\Delta x), \tag{7.2}$$

$$\Delta x' = \gamma(\Delta x - v\Delta t). \tag{7.3}$$

These equations express the components of the same spacetime displacement $\widehat{\Delta s}$ relative to different reference frames. The displacement $\widehat{\Delta s}$ connects two events with a time-like interval and its magnitude equals to the proper time ΔT_0 between these events. The proper time is a number, an invariant quantity. We can divide both sides of the Eqs. (7.2) and (7.3) by the same number ΔT_0 and obtain the rule for the *transformation of components of spacetime velocity*:

$$u_t' = \gamma(u_t - vu_x), \tag{7.4}$$

$$u_x' = \gamma(u_x - vu_t). \tag{7.5}$$

Here

$$u_t = \frac{\Delta t}{\Delta T_0}, \quad u_x = \frac{\Delta x}{\Delta T_0},$$

and

$$u'_t = \frac{\Delta t'}{\Delta T_0}, \quad u'_x = \frac{\Delta x'}{\Delta T_0},$$

are the components of *the same spacetime velocity* $\widehat{u}$ relative to different reference frames. This is an example of a more general rule, regarding spacetime vectors and Lorentz transformation:

▷ **Lorentz Transformation**

The Lorentz transformation connects the components of any spacetime vector $\widehat{\alpha} = (\alpha_t, \alpha_x)$ *in different reference frames in relative motion:*

$$\alpha'_t = \gamma(\alpha_t - v\alpha_x), \tag{7.6}$$

$$\alpha'_x = \gamma(\alpha_x - v\alpha_t). \tag{7.7}$$

? **Problem 7.5**

A particle has the spacetime velocity components

$$\widehat{u} = (1, 0)$$

in a reference frame A. Use Lorentz transformation to find the components relative to another reference frame B moving relative to A with the velocity v.

? **Problem 7.6** **

Using the Lorentz transformation only for the spatial component of the velocity

$$u'_x = \gamma(u_x - vu_t),$$

show that

$$u' = \frac{|u - v|}{1 - uv}.$$

Reminder: $u'_x = \gamma(u')u'$, $u_x = \gamma(u)u$ and $u_t = \gamma(u)$ are the spatial and temporal components of the spacetime velocity.

7.5.4 *Spacetime Momentum*

In Newtonian mechanics momentum of a body is given by the product of its mass and velocity:

$$\mathbf{p} = m\mathbf{v}.$$

The *spacetime momentum* is defined similarly:

$$\widehat{p} = m\widehat{u}, \tag{7.8}$$

where m is the mass of the particle—an invariant quantity. The components of the spacetime momentum are

$$\widehat{p} = m(\gamma, \gamma u) = (\gamma m, \gamma m u) = (p_t, p_x). \tag{7.9}$$

The spatial component $p_x = \gamma m u$ resembles Newtonian momentum multiplied by the factor γ. As discussed earlier, p_x is the relativistic momentum of a body.

The meaning of the time component p_t can be clarified if we recall the approximate expression for the factor γ:

$$\gamma \approx 1 + u^2/2,$$

therefore

$$p_t = \gamma m = m + \frac{mu^2}{2}. \tag{7.10}$$

This is the sum of the kinetic energy due to the motion of the object, and the term m present even at rest. *The time component p_t of spacetime momentum $\widehat{p}$ is the total energy of the object, including its kinetic energy.*

The relativistic expression for the energy, which we will denote E, is then given by

$$E(v) = p_t = \gamma m.$$

For an object at rest, the energy (*rest-energy*[4]) is determined by its mass:

$$E_0 = E(v = 0) = m. \tag{7.11}$$

---

[4] We can also call it *proper energy*.

The approximates value for the temporal component of the spacetime momentum can be written in meter-based units:

$$p_t = m + \frac{m\bar{u}^2}{2\bar{c}^2},$$

from which follows

$$p_t\bar{c}^2 = m\bar{c}^2 + \frac{m\bar{u}^2}{2}.$$

From this expression, it is clear that in meter-based units the expression for the rest-energy of a particle with the mass m is given by

$$\bar{E}_0 = m\bar{c}^2.$$

(The units of mass are the same in meter-based and c-based units).

The relativistic expression for momentum is

$$P = p_x = \gamma m u.$$

Note that the relativistic energy and relativistic momentum are related in a simple way:

$$P = \gamma m u = Eu.$$

The spacetime momentum can be written in any of the following ways:

$$\widehat{p} = (E, P) = (\gamma m, \gamma m u) = (p_t, p_x). \tag{7.12}$$

The conservation of relativistic energy E and momentum P is an experimentally established fact. Countless experiments, involving collision of elementary particles moving at high speeds ($u \approx 1$), confirm the conservation of both relativistic energy E and momentum P. The conclusion is the following:

▷ **Spacetime Momentum Is Conserved**

Spacetime momentum

$$\widehat{p} = (E, P)$$

is conserved. Its time component $p_t = E$ is conserved, and its space component $p_x = P$ is conserved.

Later we will see how to apply this conservation law to the analysis of collisions and decays of fast-moving objects.

Spacetime momentum is defined as mass multiplied by a spacetime velocity. The former is an invariant quantity, and the latter is a spacetime vector. Therefore, the momentum $\widehat{p} = m\widehat{u}$ is also a spacetime vector. It points in the same direction as the spacetime velocity $\widehat{u}$, and its length squared is proportional to the lengths squared of the velocity:

$$p^2 = m^2 u^2 = m^2.$$

In terms of spacetime components

$$p_t = mu_t, \quad p_x = mu_x,$$

the length squared of $\widehat{p}$ can be written as follows:

$$p^2 = m^2 u_t^2 - m^2 u_x^2 = p_t^2 - p_x^2.$$

It can also be written in terms of E and P:

$$p^2 = E^2 - P^2.$$

Combining the last formula with the fact that $p^2 = m^2$, we arrive at an important relation between the energy and momentum in the special theory of relativity:

$$E^2 - P^2 = m^2. \tag{7.13}$$

This can be written in another way

$$E^2 = m^2 + P^2. \tag{7.14}$$

When the relativistic momentum $P = \gamma mu$ (the spatial part of $\widehat{p}$) is much larger than the mass ($P \gg m$), the term m^2 can be neglected in the last equation, and the energy becomes approximately equal to the magnitude of the momentum

$$E \approx P.$$

For massless particles with $m = 0$ (e.g., photons), the relationship between the energy and momentum becomes

$$E = P. \tag{7.15}$$

Since the relativistic energy and momentum are related:

$$P = Eu,$$

the massless particle must be moving with the speed $u = 1$. For photons, the equality between energy and momentum is also established in the theory of electromagnetism.

When analyzing the approximate value of the time component of the space-time momentum p_t, we found that

$$p_t \bar{c}^2 \approx m\bar{c}^2 + \frac{m\bar{u}^2}{2}.$$

This relationship indicates that the relativistic energy E can also be written in the meter-based units:

$$\bar{E} = E\bar{c}^2 \approx m\bar{c}^2 + \frac{m\bar{u}^2}{2}.$$

Similar analysis of the spatial component of the spacetime momentum shows that

$$p_x \bar{c} \approx m\bar{v},$$

which implies

$$\bar{P} = P\bar{c}$$

as the expression of the relativistic momentum in meter-based units. Given that

$$E = \frac{\bar{E}}{\bar{c}^2}, \quad P = \frac{\bar{P}}{\bar{c}},$$

the relationship between the relativistic energy, momentum, and mass takes the form

$$\frac{\bar{E}^2}{\bar{c}^4} = m^2 + \frac{\bar{P}^2}{\bar{c}^2}$$

or, equivalently,

(continued)

$$\bar{E}^2 = m^2\bar{c}^4 + \bar{P}^2\bar{c}^2.$$

In c-based units, this relationship looks neater:

$$E^2 = m^2 + P^2.$$

? Problem 7.7

A particle is at rest in a given reference frame S. Its spacetime momentum has the components

$$\widehat{p} = (m, 0).$$

Apply Lorentz transformation to find the components of $\widehat{p}$ relative to a reference frame S', moving with the speed v relative to S.

7.5.5 *Spacetime Acceleration**

Similarly to spacetime velocity, we can define *spacetime acceleration*:

$$\widehat{a} = \frac{\Delta\widehat{u}}{\Delta T_0}; \tag{7.16}$$

it tells how the spacetime velocity $\widehat{u}$ changes with the proper time, measured along a worldline. Spacetime acceleration is related to the change of the "direction" or slope of a worldline. The worldlines of inertial observers are straight—the spacetime velocity always points in the same direction—resulting in zero spacetime acceleration. Figure 7.12 shows three inertial observers: A, B, and C; it also shows a curved worldline of an accelerating object.

Later we will study in details a specific example of a curved worldline, describing the motion of an accelerating objects. We will not use accelerated reference frames, but will simply study an object accelerating relative to an inertial frame.

7.6   Energy-Mass Equivalence

Mass of any object is a relativistic invariant—it is the same for all inertial observers. This, however, does not mean that it always remains constant; mass of a body can change even in a given reference frame. For example, when an electron and positron collide, they annihilate, producing high-energy electromagnetic radiation. The reaction is symbolically written as follows:

$$e^- + e^+ \rightarrow \gamma + \gamma,$$

where γ represents a high energy photon. The total mass before the annihilation is $2m_e$, whereas after the annihilation two massless particles appear. The rest-energies of the electron and the positron are converted into the kinetic energy of photons. A reverse process is also possible, where the kinetic energy of a photon can be converted into the rest-energy of two massive particles:

$$\gamma \rightarrow e^- + e^+.$$

The conservation of relativistic momentum requires energy and mass to be connected. To demonstrate this, let's examine an *inelastic collision* of two particles, illustrated in Fig. 7.13.

In a reference frame S two identical particles are moving towards each other with equal speeds U along the y axis. The mass of each particle is m. After the inelastic collision they stop and form a new particle with mass M. The parts (a) and (b) of Fig. 7.13 show this process in the reference frame S.

The reference frame S' is moving relative to S with the speed v in the positive direction of the x axis. In S', the colliding particles and the resulting particle are moving with the velocity v in the negative direction of the x' axis, as shown in Fig. 7.13c, d. Using the velocity composition formulas, we can find the components of the velocities along the y' axis:

$$U/\gamma(v)$$

for the particle at the bottom, and

$$-U/\gamma(v)$$

for the particle on the top.

The conservation of momentum along the x' axis requires that

$$2P_m(w) \sin \alpha = P_M,$$

where $P_m(w)$ is the relativistic momentum of each original particle; $P_M(v)$ is the relativistic momentum of the resulting particle. Using the expression for the

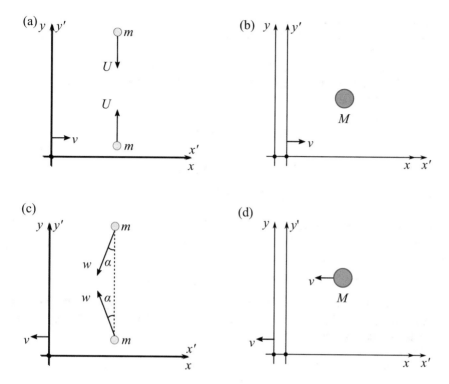

Fig. 7.13 Two identical particles, each with mass m, collide inelastically and form a new single particle with mass $M > 2m$. (a), (b): In the reference frame $S = (x, y)$, the particles are moving along the y axis and the resulting particle is at rest. (c), (d): In the reference frame $S' = (x', y')$, moving with the speed v relative to S, the particles before the collision and the final particle have the components of momentum along the x' axis

relativistic momentum, we can rewrite the conservation requirement as

$$2\gamma(w)mw\sin\alpha = \gamma(v)Mv,$$

where

$$w = \sqrt{v^2 + U^2/\gamma^2(v)} \quad (w > v)$$

is the speed of each particle before the collision in the reference frame S'. Given that

$$w\sin\alpha = v,$$

we obtain

$$M = 2\frac{\gamma(w)}{\gamma(v)}m.$$

Clearly, $w > v$ and therefore $\gamma(w) > \gamma(v)$, meaning that the mass M of the resulting particle will be greater than the sum of the masses of two original particles. Where does the extra mass come from?

Consider the case of small relative speed of the reference frames ($v \ll 1$), and moderately fast moving particles ($U \gg v$). Under these conditions, we can approximately write

$$\gamma(v) \approx 1, \quad w \approx U;$$

the mass of the resulting particle becomes

$$M = 2\gamma(U)m.$$

Using the approximation for $\gamma(U)$

$$\gamma(U) \approx 1 + \frac{U^2}{2},$$

it becomes clear that the rest-energy (mass) of the resulting particle has the contributions from the rest-energies (masses) of the original particles, as well as from their kinetic energy:

$$M \approx 2m + 2\frac{mU^2}{2}.$$

When the speed U of the colliding particles is comparable to the speed of light, the change of the total mass of the system is given by

$$\Delta M = M - 2m = 2m(\gamma - 1).$$

This change of the mass, as we saw above, is coming from the kinetic energy. The relativistic expression for the kinetic energy of a particle moving with speed U is then

$$E_k = m[\gamma(U) - 1].$$

It equals zero at $U = 0$, then approximately equals to $mU^2/2$, as in Newtonian mechanics, and finally begins to increase rapidly, as U approaches the speed of light (see Fig. 1.8). From the relativistic expression for the kinetic energy follows that *an infinite amount of energy is required to accelerate a massive particle to the speed of light.*

In Newtonian mechanics the kinetic energy in inelastic collisions "disappears". It may be converted into heat, sound, and other forms of energy, without affecting mass. The special theory of relativity establishes an intimate connection between energy and mass. In the words of Albert Einstein[5]

> ▷ **Einstein on Energy-Mass Equivalence**
>
> *It followed from the special theory of relativity that mass and energy are both but different manifestations of the same thing—a somewhat unfamiliar conception for the average mind. Furthermore, the equation E is equal to m c-squared, in which energy is put equal to mass, multiplied by the square of the velocity of light, showed that very small amounts of mass may be converted into a very large amount of energy and vice versa. The mass and energy were in fact equivalent, according to the formula mentioned above.*

7.7 Practice in Energy-Momentum

To see how the conservation of spacetime momentum is applied, it is useful to study a couple of examples.

First, consider a stationary particle with mass M. It decays into two identical particles, each having mass $m < M/2$. The situation is illustrated in Fig. 7.14. Let's find the speed of each product particle.

The conservation of energy requires that the total energy before the decay equals the total energy after. Before the decay we have the rest-energy of the particle with mass M. After the decay we have two particles moving in opposite directions with equal momenta and speeds (to conserve momentum). The total energy (rest energy plus kinetic energy) of a moving particle with mass m is

$$E = \gamma m,$$

Fig. 7.14 A particle with mass M is at rest. It then decays into two identical product particles, each with mass m. Find the speed of each product particle

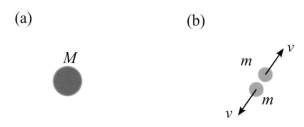

(a)

(b)

---

[5] From the soundtrack of the 1948 film, *Atomic Physics*. See also *Einstein Explains the Equivalence of Energy and Matter* at https://history.aip.org/exhibits/einstein/voice1.htm.

and the conservation of energy takes the form

$$M = 2\gamma m.$$

From the last expression we find

$$\gamma = \frac{M}{2m},$$

from which follows

$$v = \sqrt{1 - \frac{1}{\gamma^2}} = \sqrt{1 - \frac{4m^2}{M^2}}.$$

As the mass of the product particle becomes smaller, its speed becomes greater. In the limit of $m \rightarrow 0$ (massless particles, e.g. photons) the product particles must move with the speed of light.

Next, let us consider a more complicated case, where the original particle with the mass M is moving with the speed v. The particle decays into two subparticles with equal masses m. The product particles move symmetrically relative to the trajectory of the original particle, as shown in Fig. 7.15. Let's find the angle between the trajectories of the product particles.

The conservation of energy requires that

$$\gamma(v)M = \gamma(u)m + \gamma(w)m.$$

Here we allowed the speeds of the particles to be different. This equation does not involve any angles, therefore we need to use more information.

The conservation of momentum can be written for the x axis:

$$\gamma(v)Mv = \gamma(u)mu \cos\alpha + \gamma(w)mw \cos\alpha,$$

(a)   (b)

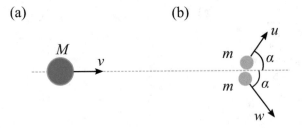

Fig. 7.15 A particle with mass M is moving with speed v. It then decays into two identical product particles which move symmetrically relative to the direction of motion of the original particle. Find the angle between the velocities of the product particles, given their equal masses m

and the y axis:

$$0 = \gamma(u)mu \sin\alpha - \gamma(w)mw \sin\alpha.$$

From the last equation follows that $\gamma(u)u = \gamma(w)w$, which implies $u = w$. Using the equality $u = w$, the conservation of energy simplifies to

$$\gamma(v)M = 2\gamma(u)m,$$

from which it is possible to find $\gamma(u)$ and u:

$$\gamma(u) = \frac{\gamma(v)M}{2m},$$

$$u = \sqrt{1 - \frac{4m^2}{M^2\gamma^2(v)}} = \sqrt{1 - \frac{4m^2}{M^2}(1 - v^2)}.$$

Finally, the conservation of momentum along the x axis takes the form

$$\gamma(v)Mv = 2\gamma(u)mu \cos\alpha,$$

from which follows:

$$\cos\alpha = \frac{\gamma(v)Mv}{2\gamma(u)mu} = \frac{v}{u} = \frac{v}{\sqrt{1 - \frac{4m^2}{M^2}(1 - v^2)}}.$$

Once the angle α is found, we know the total angle $\theta_{total} = 2\alpha$.

As a special case, let's examine what happens when the original particle is at rest ($v = 0$). Then $\cos\alpha = 0$, meaning $\alpha = \pi/2$, and the product particles move in the opposite directions. Their speeds are given by

$$u = \sqrt{1 - \frac{4m^2}{M^2}} \; ;$$

it the same expression we found earlier, having solved the problem with a stationary particle.

* * *

Below are several problems for the reader to solve, in order to get more practice working with the relativistic energy and momentum.

Fig. 7.16 For Problem 7.8:
An atom at rest absorbs a
photon and changes its mass

Fig. 7.17 For Problem 7.9:
An atom at rest emits a
photon and recoils

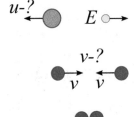

Fig. 7.18 For Problem 7.10:
Two protons with equal
speeds collide head-on and
create another particle.
Protons do not disappear; all
particles are at rest after the
collision

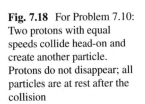

? Problem 7.8

An atom with the mass M is at rest. It absorbs a photon with the energy E. Find
the energy and momentum of the atom after the absorption of the photon. Find the
mass of the atom after the absorption (Fig. 7.16).

? Problem 7.9

An atom with the mass M is at rest. It emits a photon with the energy E. Find the
speed of the atom after the emission (Fig. 7.17).

? Problem 7.10

Two protons are colliding head-on with equal speeds. As the result of the collision,
a third particle Y is created, having the mass M. The protons and the particle Y are
all at rest. Find the speeds of the protons before the collision. Assume that the mass
of a proton is a known constant m_p (Fig. 7.18).

Fig. 7.19 For Problem 7.11: An unstable particle with mass M is at rest. It then decays into two photons

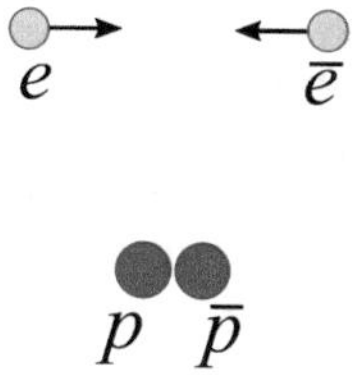

Fig. 7.20 For Problem 7.12: The electron-positron pair creates a proton and anti-proton pair in a head-on collision

Fig. 7.21 For Problem 7.13: A moving unstable particle decays into two photons. Given the angle between the photons' momenta α, find the speed of the particle v

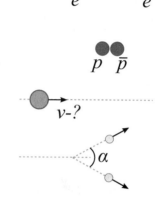

? Problem 7.11

A particle with the mass M is at rest. It decays into two photons. Find the momenta of each photon (Fig. 7.19).

? Problem 7.12

An electron collides head-on with an anti-electron (positron), both traveling at the same speed. What should be their *minimal* kinetic energy to create a pair of proton and anti-proton. Remember that kinetic energy is $E_k = E - E_0$—the relativistic energy minus the energy at rest (Fig. 7.20).

? Problem 7.13

A particle is moving left to right. It then decays into two photons. The angle between the trajectories of the photons is α. What was the speed of the particle? (Fig. 7.21).

Fig. 7.22 For Problem 7.14:
A moving particle with mass
M is emitting a photon and
stops due to recoil. Find the
initial speed of the particle

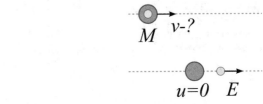

Fig. 7.23 A photon with the
energy E is scattered by an
electron at rest. The scattered
photon is moving at an angle
α relative to its original
direction and has the energy
$E' < E$

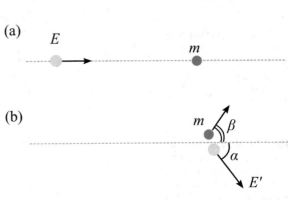

? Problem 7.14

A particle with the mass M is moving left to right. It emits a photon with the energy
E and stops. What was the initial speed of the particle? (Fig. 7.22).

7.7.1 Compton Scattering*

The conservation of energy and momentum can be used to analyze an important
physical effect of scattering of electromagnetic radiation by electrons. Electrons are
elementary particles without internal structure; as the result, electrons can not absorb
photons, only scatter them in different directions.

An American physicist Arthur Holly Compton studied scattering of X-rays by
electrons. For his contribution to physics, in particular for the study of what is now
called the Compton scattering, he received Nobel prize in 1927. Compton scattering
finds applications in various fields, ranging from radio-biology to astrophysics.

Figure 7.23 illustrates the Compton scattering: A photon with the energy E is
scattered by an electron at rest; after the scattering, the electron is moving with the
speed v at an angle β relative to the original direction of the photon. The scattered
photon is moving at an angle α. Given the initial energy of the photon, E, the mass of
the electron, m, and the angle α, we can find the energy of the scattered photon, E'.

The conservation of energy requires that

$$E + m = \gamma m + E'.$$

The horizontal component of momentum is conserved:

$$E = \gamma m v \cos \beta + E' \cos \alpha,$$

where we used the fact that for massless particles $P = E$. The conservation of the vertical component of the momentum is written as

$$\gamma m v \sin \beta = E' \sin \alpha.$$

From the last two equations it is possible to find the magnitude of the electron's spatial momentum squared:

$$\gamma^2 m^2 v^2 = (E - E' \cos \alpha)^2 + (E' \sin \alpha)^2.$$

Opening the parentheses on the right-hand side and simplifying, we get

$$\gamma^2 m^2 v^2 = E^2 + E'^2 - 2EE' \cos \alpha.$$

Recalling that $\gamma^2 v^2 = \gamma^2 - 1$, we can transform the last equation:

$$\gamma^2 m^2 = m^2 + E^2 + E'^2 - 2EE' \cos \alpha. \tag{7.17}$$

The reason for this transformation is that the left-hand side can be obtained from the conservation of energy:

$$\gamma m = m + E - E',$$

consequently,

$$\gamma^2 m^2 = m^2 + E^2 + E'^2 + 2mE - 2mE' - 2EE'. \tag{7.18}$$

Combining the Eqs. (7.17) and (7.18), after some simplifications, we find the desired expression:

$$E' = \frac{mE}{m + E(1 - \cos \alpha)}.$$

The last equation is sometimes written in a slightly different way:

$$E'(\alpha) = \frac{E}{1 + (1 - \cos \alpha)E/m}. \tag{7.19}$$

The photon transfers maximum energy to the electron when the final energy E' has the lowest possible value. It happens for $\alpha = \pi$, meaning that the photon is scattered in the direction opposite to the original motion.

Sunyaev-Zeldovich Effect

In the process of Compton scattering a photon can not only loose energy, but also gain energy. For this to happen, the scenario considered above must unfold "in reverse". This is called the *inverse Compton scattering*—an effect important in astrophysics.

In 1969 astrophysicists Rashid Sunyaev and Yakov Zeldovich predicted that when photons from the Cosmic Microwave Background (CMB) radiation[6] are scattered from fast electrons in an inverse Compton effect, they will gain energy and the resulting spectrum of CMB from that region will change. Fast electrons are usually present in galaxy clusters. Studying the deviations of CMB spectrum from its normal shape, astrophysicists can infer various information about the astronomical objects and processes involving fast electrons. For example, it is possible to estimate the diameter of a galaxy cluster this way.

7.8   Accelerated Motion*

According to the second law of Newtonian mechanics, non-zero net force applied to an object creates acceleration proportional to the force. Acceleration may result in the change of the magnitude of the velocity, its direction, or both.

In an important case of circular motion, illustrated in Fig. 7.24, an object can keep its speed constant and yet move with an acceleration—*centripetal acceleration*. The magnitude of this acceleration is given by

$$a_C = \frac{v^2}{R}.$$

During the circular motion, the distance from the center of the circle to the object remains constant:

$$x^2 + y^2 = R^2.$$

The motion we consider next is, in some sense, analogous to circular motion: An object starts d seconds away from the origin at time $t = 0$, according to an observer A (see Fig. 7.25); it then starts accelerating while *keeping the spacetime distance OE constant*. In the reference frame A this requirement is written as

---

[6] CMB is an electromagnetic radiation present everywhere in the observable universe. Its spectrum is very broad, resembling the electromagnetic radiation of a very cold body, with the temperature of $-270.4°$ Celsius.

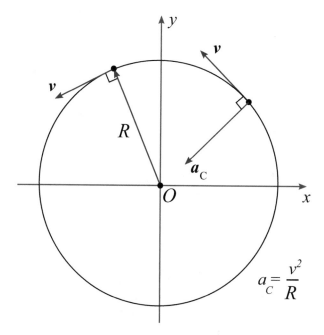

Fig. 7.24 In a circular motion with constant speed, the direction of the motion is changing. The acceleration is always pointing perpendicular to the velocity, towards the center of the circle

$$t^2 - x^2 = -d^2 = \text{const.} \qquad (7.20)$$

The minus sign in front of d^2 is due to the fact that the interval $\widehat{s}$ is always space-like.

At some later time $t + \delta t$ the object's position will change to $x + \delta x$. Since the spacetime interval is required to remain the same, we get[7]

$$(t + \delta t)^2 - (x + \delta x)^2 = -d^2. \qquad (7.21)$$

The difference between the last two expressions is

$$\left[(t + \delta t)^2 - t^2\right] - \left[(x + \delta x)^2 - x^2\right] = 0. \qquad (7.22)$$

Using the equality

$$a^2 - b^2 = (a - b)(a + b),$$

---

[7] Refer to the Appendix A.2 for the meaning and use of δ notation.

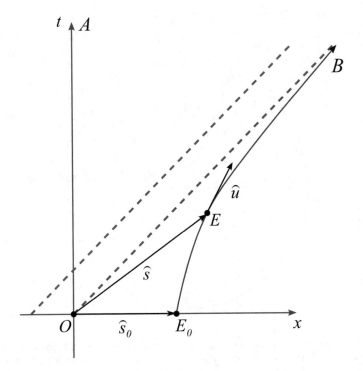

Fig. 7.25 l diagram for an object moving with changing velocity, while keeping the separation Δs from the origin O constant

we can rewrite the previous equation:

$$\delta t(2t + \delta t) - \delta x(2x + \delta x) = 0, \tag{7.23}$$

from which follows the coordinate velocity

$$v = \frac{\delta x}{\delta t} = \frac{t + \delta t/2}{x + \delta x/2}. \tag{7.24}$$

For a vanishingly small time interval δt, the displacement $\delta x = v\delta t$ will also be vanishingly small, and the formula for the coordinate velocity reduces to

$$v = \frac{t}{x}. \tag{7.25}$$

(Note that this formula differs from the case of the motion with constant speed, when $v = x/t$.)

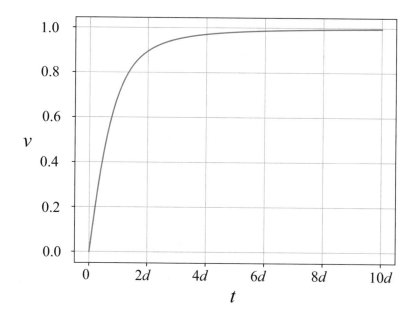

Fig. 7.26 Time dependence of the speed of the object moving along the worldline of constant interval $\Delta s^2 = -d^2$. An object is constantly accelerating, but never reaches the speed of light

The γ-factor, corresponding to (7.25), is

$$\gamma = \frac{1}{\sqrt{1 - t^2/x^2}} = \frac{x}{d}. \qquad (7.26)$$

Using the constant spacetime separation, we can write the time dependence of the spacetime position

$$\widehat{s} = (t, x) = (t, \sqrt{d^2 + t^2}),$$

and the object's coordinate velocity

$$v = \frac{\delta x}{\delta t} = \frac{t}{\sqrt{d^2 + t^2}}. \qquad (7.27)$$

For a long enough time, the object's speed approaches the speed of light, but never reaches it, as shown in Fig. 7.26. The spacetime velocity of the object is

$$\widehat{v} = (\gamma, \gamma v) = \frac{1}{d}(x, t). \qquad (7.28)$$

Interestingly, as shown in Fig. 7.25, the non-inertial observer B will not be able to receive light signals from any event lying in the negative part of the axis x.

What about the acceleration of the object? It is easily found from the definition

$$\widehat{a} = \frac{\delta \widehat{v}}{\delta T_0} = \gamma \frac{\delta \widehat{v}}{\delta t}. \tag{7.29}$$

Using the components of the spacetime velocity, we get

$$\delta \widehat{v} = \frac{1}{d}(\delta x, \delta t), \tag{7.30}$$

and therefore

$$\widehat{a} = \frac{\gamma}{d}(\frac{\delta x}{\delta t}, \frac{\delta t}{\delta t}) = \frac{\gamma}{d}(\frac{t}{x}, 1). \tag{7.31}$$

Recalling that $\gamma = x/d$, the expression for the spacetime acceleration can be reduced to

$$\widehat{a} = \frac{1}{d^2}\widehat{s}.$$

The spacetime acceleration is proportional to the spacetime position; since the latter has constant magnitude (we study the motion with constant spacetime interval $\Delta \widehat{s}$), the spacetime acceleration also has constant magnitude; the magnitude squared given by

$$a^2 = \frac{s^2}{d^4} = -\frac{1}{d^2}.$$

Compare this to the centripetal acceleration, for the case of speed $v = 1$ and the circle of the radius $R = d$:

$$a_c^2 = \frac{1}{d^2}.$$

It is convenient to rewrite the expressions for the coordinate velocity in terms of time t and the magnitude of the acceleration a:

$$v = \frac{at}{\sqrt{1 + (at)^2}}.$$

For times shortly after $t = 0$ the coordinate velocity v is approximately given by

$$v = at,$$

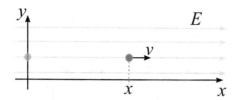

Fig. 7.27 A particle with an electric charge q is moving in a uniform electric field. The strength of the field is E. The work $W = qEx$, done by the field on the particle, is converted into the energy $\gamma(v)m$ of the latter

it corresponds to the uniformly accelerated motion with the acceleration a.

The motion discussed above can describe a particle (e.g., an electron) in a uniform electric field with the strength E, as shown in Fig. 7.27. Electric force applied to a charge q is $F = qE$. Therefore, having moved the particle by a distance $x - x_0$, the field performs work $W = F(x - x_0) = qE(x - x_0)$, which is transferred to the kinetic energy of the particle:

$$(\gamma(v) - 1)\,m = qE(x - x_0).$$

From the last expression we find

$$\gamma(v) = \frac{qE}{m}(x - x_0) + 1 = \frac{qE}{m}x + (1 - \frac{qE}{m}x_0).$$

Comparing this to the expression for γ in (7.26), we can see that their dependencies on the position x are very similar: in both cases γ grows linearly with the coordinate x. If we choose the electric field strength such that

$$\frac{qE}{m} = \frac{1}{d},$$

and the initial position (i.e., the placement of the origin of the coordinate axis x)

$$x_0 = \frac{m}{qE},$$

the dependence of the γ factor becomes

$$\gamma = \frac{x}{d}.$$

We can conclude that the motion of the electron in this uniform electric field will be described by the formulas of accelerated motion with constant spacetime interval, derived above.

Chapter Highlights

- Although the relative motion of the reference frames S and S' along the axis x does not affect the measurements of the coordinates $y = y'$ and $z = z'$, it affects the measurements of the velocities along those axes.
- Newtonian momentum $\mathbf{p} = m\mathbf{v}$ is not conserved in particle collisions.
- Relativistic expression for momentum is given by $P = mv/\sqrt{1 - v^2}$.
- Mass of an object does not depend on its velocity. Mass is a relativistic invariant.
- Relativistic momentum P is the spatial component of spacetime vector—spacetime momentum. The temporal component of spacetime momentum is relativistic energy $E = \gamma m$.
- The magnitude of the spacetime momentum is an invariant quantity; it equals to the mass of an object.
- Lorentz transformation, generally speaking, relates the components of various spacetime vectors, as measured by different observers in relative motion.
- Energy and mass are equivalent: Mass (rest-energy) of an object can be transformed into energy, and vice versa.
- Although accelerated observers are not inertial, and thus are not used in the special theory of relativity, the theory itself can study the motion of accelerating objects.

References

1. Russell, B. (1977). *ABC of Relativity* (4th ed.) London: Allen and Unwin.
2. Okun, L. B. (1989). The Concept of Mass. *Physics Today, 42*(6), 31–36.
3. Taylor, E., & Wheeler, A. (1992). *Spacetime Physics* (2nd ed., p. 78). New York: Freeman and Company.
4. *Einstein Explains the Equivalence of Energy and Matter.* https://history.aip.org/exhibits/einstein/voice1.htm.

Closing Remarks

> *We shall not cease from exploration*
> *And at the end of all our exploring*
> *Will be to arrive where we started*
> *And know the place for the first time*

T.S. Eliot *Little Gidding,* Four Quartets 1943

What We Left Out?

In the previous chapters we learned the basic ideas and mathematical tools of the special theory of relativity. The subject, understandably, has much more to offer. Two big topics were not discussed: The formulation of relativistic mechanics, including the concepts of force and work; and the theory of electricity and magnetism. Beyond that, an advanced learner may be curious about the formulation of the theory for the case of general relative motion of two observers, not confined to the motion along the x axis. Finally, a very broad topic of how the special theory of relativity fits in the rest of the physics, including thermodynamics, hydrodynamics, field theories, and particle physics.

A very different book, or a set of books, is required to properly cover the left-out topics. Furthermore, a significantly different level of preparation and devoted time is expected from anyone who pursues those interesting questions. It is hoped that this book can give both the initial momentum and energy to explore the rest of the fascinating areas of relativity, and physics in general.

Suggested Reading

The special theory of relativity is more than a century old and there many books written about it. There are short books and long books, popular and sophisticated, technical and philosophical. It is easy to get lost in the plethora of the material available on the subject. Below is a short list of recommended books about the theory of relativity. The books with an asterisk (e.g., H. Stephani, *Relativity***) are

© The Author(s), under exclusive license to Springer Nature Switzerland AG 2022
Y. Deshko, *Special Relativity*, Undergraduate Lecture Notes in Physics,
https://doi.org/10.1007/978-3-030-91142-3

more advanced and mathematically challenging; the more asterisks—the greater the challenge. The books without the asterisk should be even more accessible than the current book.

1. E. F. Taylor, J. A. Wheeler, *Spacetime Physics*, 2nd ed. W. H. Freeman and Company, New York, 1991.
2. A. Einstein, *Relativity: The Special and General Theory*, Dover, 2010.
3. B. Russell, *ABC Of Relativity*, 4th ed. Allen and Unwin, London, 1977.
4. H. Bondi, *Relativity and Common Sense*, Anchor Books, New York, 1964.
5. J. L. Synge, *Talking About Relativity*, Elsevier, New York, 1970.
6. L. D. Landau, G. B. Rumer, *What is Relativity*, Dover, New York, 2003.
7. M. Born, *Einstein's Theory of Relativity**, Dover, 1962.
8. H. Stephani, *Relativity***, Cambridge University Press, 2004.

Appendix A
Mathematical

A.1 Geometric Series

Given a real number q, such that $q < 1$, we can calculate the infinite sum

$$S = 1 + q + q^2 + q^3 + \ldots + q^n + \ldots$$

First, we can write the sum with a finite number of terms:

$$S_n = 1 + q + q^2 + q^3 + \ldots + q^n,$$

and

$$S_{n+1} = 1 + q + q^2 + q^3 + \ldots + q^n + q^{n+1}.$$

For large enough n, the difference between S_n and S_{n+1} can be arbitrarily small. The relationship between S_n and S_{n+1} can be found, if we write

$$S_{n+1} = 1 + q(1 + q + q^2 + \ldots + q^{n-1} + q^n) = 1 + q S_n.$$

When n is infinitely large, both S_n and S_{n+1} approach the same value $S_\infty = S$. Therefore,

$$S = 1 + qS,$$

from which follows

$$S = \frac{1}{1-q} = 1 + q + q^2 + q^3 + \ldots + q^n + \ldots.$$

© The Author(s), under exclusive license to Springer Nature Switzerland AG 2022
Y. Deshko, *Special Relativity*, Undergraduate Lecture Notes in Physics,
https://doi.org/10.1007/978-3-030-91142-3

For the values of $q \ll 1$, the approximation

$$\frac{1}{1-q} \approx 1 + q$$

works well. For example, for $q = 0.1$, we have

$$\frac{1}{1-q} = 1.1(1),$$

and

$$1 + q = 1.1(0).$$

A.2 $\Delta - \delta$-notation

Δ notation is used when discussing a difference or a change of a quantity, and this change can be of any magnitude. For example: Given two points P and Q in a plane, we may find their coordinates in a Cartesian coordinate system $S = \{O, x, y\}$. The point P gets the coordinates (x_P, y_P), the point Q gets the coordinates (x_Q, y_Q). If we are interested in how far apart these points are along the x axis, we will calculate the difference of their x coordinates:

$$\Delta x = x_Q - x_P;$$

similarly for the y coordinate:

$$\Delta y = y_Q - y_P.$$

These differences can be quite large, if the points P and Q are far apart.

Next, imagine that we are tracking the position of an object using polar coordinates (r, ϕ). At the moment of time t_1 the object is at a point $1 = (r_1, \phi_1)$, then, at some later moment of time t_2, it is at a different point $2 = (r_2, \phi_2)$. We can be interested in the change in time

$$\Delta t = t_2 - t_1,$$

or in the change

$$\Delta r = r_2 - r_1$$

in the distance to the pole O, or in the change of the angle

$$\Delta \phi = \phi_2 - \phi_1.$$

When the change of a variable is small (sufficiently small for a given problem), then we use δ—the lowercase Greek letter delta. Using the examples introduced above, we would write

$$\delta x = x_Q - x_P,$$

if the points P and Q are very close to each other. We would also write

$$\delta t = t_2 - t_1,$$

if we measure the coordinates of the moving object in two close moments of time. Similarly, we would write

$$\delta r = r_2 - r_1$$

and

$$\delta\phi = \phi_2 - \phi_1,$$

if the distance and the angle did not change significantly over the small time interval δt.

A.3 Factor γ

Many relativistic formulas contain the factor

$$\gamma = \frac{1}{\sqrt{1 - v^2}}. \tag{A.1}$$

Here we study the behavior of this factor as the function of relative velocity v. Figure A.1 shows the values of γ at various speeds v, including a focus on the speeds below $1/2$ of the speed of light.

At low relative speeds the curve $\gamma(v)$ resembles a parabola. To find the approximate expression for γ in the limit $v \ll 1$, first observe that $\gamma > 1$ always, and $\gamma \approx 1$ for $v \ll 1$. Thus, we can write

$$\gamma \approx 1 + \epsilon, \quad \epsilon \ll 1.$$

From (A.1) we get

$$\gamma^2 = \frac{1}{1 - v^2} \approx 1 + v^2.$$

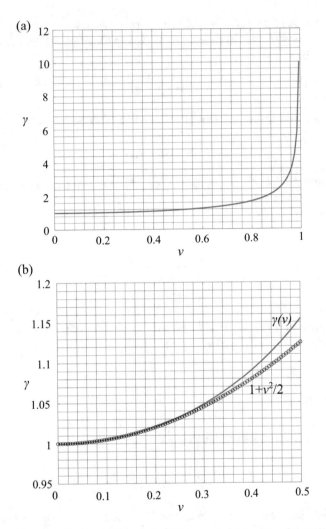

Fig. A.1 Behavior of the relativistic factor γ for different values of relative speed v. (a) The value of γ explodes near $v \approx 1$. (b) For small and moderate speed ($v < 0.3$) the curve $\gamma(v)$ is parabolic

Here we used the approximation formula for the geometric series. From the last expression follows

$$v^2 = \gamma^2 - 1.$$

Using the algebraic identity $(a - b)(a + b) = a^2 - b^2$, we can write

$$v^2 \approx (\gamma - 1)(\gamma + 1) \approx 2\epsilon + \epsilon^2.$$

Given that $\epsilon \ll 1$, we can neglect the ϵ^2 term, to obtain

$$\epsilon = \frac{v^2}{2},$$

from which follows

$$\gamma = 1 + \frac{v^2}{2}.$$

The bottom part of Fig. A.1 shows how well this quadratic formula (black open circles) approximates the exact expression for speeds below 0.2.

Appendix B
Physical

B.1 Symmetric Elastic Collision

The problem discussed below is neither difficult nor particularly important in Galileo-Newtonian mechanics. However, it provides a simple way to demonstrate the issue with the non-relativistic expression for the momentum

$$p = mv.$$

In addition, analyzing this problem with the help of relativistic velocity transformation formulas, we can arrive at the relativistic expression for the momentum

$$P = \gamma mv.$$

This is why the problem is discussed in details. The analysis is carried out from the point of view of the Galileo-Newtonian mechanics only to make the situation intuitively clear.

Observers S and S' are standing on different platforms that are moving relative to each other with the speed v along the parallel rails, as shown in Fig. B.1b. The observers have identical elastic particles, which they throw with the speed U perpendicular to the rails, as viewed in their frames of reference. The particles collide *symmetrically*: they bounce off each other and return back to the observers. We need to find the velocity of each particle after the collision. *An important note*: We will neglect the effects of gravity on the motion of the elastic particles. This can be done if the experiment is carried out in a low-gravity environment (space), or if the collision happens so fast that the motion due to gravity does not have time to play a role.

It is worth elaborating on some details of this problem, as it may appear that the required collision is difficult to achieve. Let us start with the situation shown in Fig. B.1a, when the rails are separated by a solid immovable wall with the thickness

© The Author(s), under exclusive license to Springer Nature Switzerland AG 2022
Y. Deshko, *Special Relativity*, Undergraduate Lecture Notes in Physics,
https://doi.org/10.1007/978-3-030-91142-3

Fig. B.1 (a) Two observers
are throwing particles towards
an immovable wall,
independently of each other.
It is possible to choose their
positions in such a way that
the particles hit the wall at the
same time at the opposite
sides. (b) Once the
simultaneous collision of the
particles with the wall is
achieved, we can imagine the
wall becoming very thin and,
eventually, disappear. The
result is a symmetric collision
between the particles

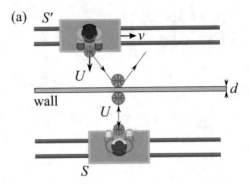

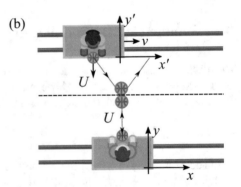

d. The observer S is at rest relative to the wall; an elastic particle (a rubber ball, for example) is thrown with the speed U perpendicular to the rails and the wall. It bounces off the wall and travels back into its original position.

On the other side of the wall, the observer S' is moving along the rails with the speed v relative to the wall. An elastic particle, identical to the one carried by the observer S, is thrown by the observer S' towards the wall. In the reference frame of S', the particle is moving perpendicular to the rails and the wall with the speed U. The particle of the observer S' will bounce off the wall and travel back to its original position, always moving perpendicular to the wall from the point of view of S'. Both S and S' observe the same behavior of their respective particles.[1]

In the reference frame of the wall (and the observer S), the elastic particle of the observer S' will be moving along the rails with the speed v, in addition to the motion towards the wall. This particle will hit the wall at a certain point and moment of time. By appropriately choosing when and where the observer S emits its particle, we can make both particles collide with the wall at the same time, on exactly opposite sides

---

[1] For this to happen, strictly speaking, there must be no friction between the balls and the wall.

Fig. B.2 Symmetric elastic
collision of two "particles".
(**a**) The collision observed
from the reference frame S.
(**b**) The same collision
observed from the reference
frame S'

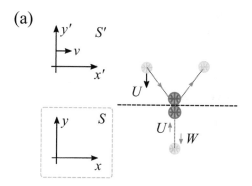

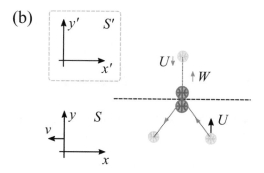

of the wall. Now we can imagine that the wall is becoming increasingly thin and,
finally, disappears, resulting in the collision of the particles, illustrated in Fig. B.1b.

To make the situation perfectly symmetric, we must take the wall and the rails
out of the picture, focusing only on the colliding particles. The final configuration
is shown in Fig. B.2.

The analysis of this collision from the point of view of Galileo-Newtonian
mechanics is straightforward. In the reference frame of S, the conservation of
momentum for the horizontal motion requires

$$mv = mv,$$

meaning that the particle of the observer S' has the same horizontal component of
the velocity before and after the collision. For the vertical motion we have

$$mU - mU = mW - mW.$$

Here, firstly, we allow the velocity of each particle before the collision to differ
from the velocity after the collision. As long as both particles have the same speeds,
moving in the opposite directions results in the total zero momentum before and

after the collision. Secondly, since the relative motion of the observers S and S' is happening along the horizontal axis, the vertical components of the particles are the same in both reference frames. Specifically, given that the particle of S' has the y' component of its velocity equal $-U$, it will have the same component along the y axis of the observer S. *This is not so in the special theory of relativity.*

Finally, the conservation of kinetic energy requires

$$\frac{mU^2}{2} + \left(\frac{mU^2}{2} + \frac{mv^2}{2}\right) = \frac{mW^2}{2} + \left(\frac{mW^2}{2} + \frac{mv^2}{2}\right).$$

The terms in the parentheses correspond to the kinetic energy of the particle from S', written relative to the reference frame of S. From the last equation follows the requirement of elastic collision:

$$U^2 = W^2 \ \rightarrow \ U = W.$$

B.2   Relativistic Momentum

Relativistic expression for the momentum of a particle with mass m, moving with the velocity v can be deduced from the analysis of symmetric collision (see Sect. B.1 above).

Firstly, let us denote the relativistic momentum using capital letter $P(v)$, and state the facts that $P(v) \neq mv$ in the following way:

$$\frac{P(v)}{mv} = f(v), \quad f(v) \neq 1,$$

where $f(v)$ is some non-constant function of particle's velocity. In other words, we write the momentum as

$$P = f(v)mv.$$

For $v \ll 1$ the results of the special theory of relativity agree with the Galileo-Newtonian physics. Therefore, when $v \ll 1$, we must recover the Newtonian momentum, which requires

$$f(v \ll 1) \approx 1.$$

In the symmetric collision, the speed of the second particle relative to S before the collision is[2]

---

[2] The subscripts "b" or "a" stand for "before" and "after" the collision, respectively.

$$u_{2b} = \sqrt{v^2 + U^2/\gamma^2(v)} = \sqrt{v^2(1 - U^2) + U^2}.$$

The speed after the collision, u_{2a}, is given by a similar equation, replacing U with W.

The total momentum along the y axis before the collision is

$$P_{yb} = P_b^{(1)} - P_b^{(2)} \sin\theta,$$

where $P_b^{(1)}$ and $P_b^{(2)}$ are the momenta of the first and the second particle before the collision. Given the relativistic expression for momentum $P(v) = f(v)mv$, we can write

$$P_{yb} = f(U)mU - f(u_{2b})mu_{2b} \sin\theta,$$

where θ is the angle the trajectory of the particle (2) makes with the x axis.

Along the x axis we have

$$P_{xb} = f(u_{2b})mu_{2b} \cos\theta.$$

After the collision the total momentum along the y axis is

$$P_{ya} = -f(W)mW + f(u_{2a})mu_{2a} \sin\phi,$$

and along the x axis:

$$P_{xa} = f(u_{2a})mu_{2a} \cos\phi.$$

Since we do not say anything about the relationship between U and W, we have to allow the scattering angles θ and ϕ to be different.

Using the fact that the particle (2) has the same velocity along the x axis before and after the collision

$$u_{2b} \cos\theta = v = u_{2a} \cos\phi,$$

together with the conservation of momentum along the x axis

$$f(u_{2b})mu_{2b} \cos\theta = f(u_{2a})mu_{2a} \cos\phi,$$

we conclude that

$$f(u_{2b}) = f(u_{2a})$$

and consequently $u_{2b} = u_{2a}$.[3] From the last equality and the equations for u_{2b} and u_{2a} follows that $W = U$.

The components of the velocity of the particle (2) along the y axis before the collision is

$$-u_{2b} \sin \theta = -\frac{U}{\gamma(v)},$$

while after the collision it equals to

$$u_{2a} \sin \phi = \frac{W}{\gamma(v)} = \frac{U}{\gamma(v)}.$$

Taking all this into account, we can write the total momentum along the y axis before the collision as

$$P_{yb} = f(U)mU - f(u_2)mU/\gamma(v),$$

and after the collision

$$P_{ya} = -f(U)mU + f(u_2)mU/\gamma(v).$$

Here we used the fact that

$$u_{2b} = u_{2a} = u_2 = \sqrt{v^2(1 - U^2) + U^2}.$$

Requiring $P_{yb} = P_{ya}$ results in the equality

$$\gamma(v)f(U) = f(u_2).$$

When the particles are slowly moving in the y direction ($U \ll 1$), while the observers S and S' are moving fast ($U \ll v < 1$), we will get

$$f(U) \approx 1, \quad u_2 \approx v.$$

From these conditions follows that

$$f(v) = \gamma(v).$$

The case of $U \ll 1$ and $v \gg U$ describes, in either S or S', the scattering of a relativistic particle on a slowly moving particle. The relativistic expression for momentum is then

$$P(v) = f(v)mv = \gamma mv.$$

---

[3] Essentially, this is the statement that when momenta of two identical objects are equal, they must be moving with equal speeds.

In conclusion, we demonstrated that

▷ **Relativistic Momentum**

The physical quantity given by the expression

$$P = \gamma m v = \frac{mv}{\sqrt{1 - v^2}}$$

is conserved in collisions. It is the expression for momentum valid in the special theory of relativity.

B.3 Penrose Diagrams

Sir Roger Penrose, a British mathematician and mathematical physicist, received a 2020 Nobel Prize in Physics for his contributions into the theoretical studies of black holes. To study the latter, Penrose proposed special coordinates which allow plotting diagrams of infinite spacetimes on a finite patch of paper. Penrose coordinates take an infinite plane and make it finite, compressing the plane along the diagonal of the rectangular coordinate system. In the theory of special relativity, diagonal lines in Minkowski diagrams correspond to the propagation of light.

We first study Penrose coordinates using transformation of Cartesian coordinates. Given the coordinates (x, y) of a point in Cartesian coordinate system, we may turn them into new coordinates using the following transformations:

$$X = \frac{1}{\sqrt{2}} \left(\arctan\left(\frac{x+y}{\sqrt{2}}\right) + \arctan\left(\frac{x-y}{\sqrt{2}}\right) \right) ,$$

$$Y = \frac{1}{\sqrt{2}} \left(\arctan\left(\frac{x+y}{\sqrt{2}}\right) - \arctan\left(\frac{x-y}{\sqrt{2}}\right) \right) . \tag{B.1}$$

Despite their complicated look, Penrose transformation consists of three simple steps, as illustrated in Figs. B.3 and B.4.

First, we rotate the Cartesian axes by $\pi/4$ clock-wise to get

$$x' = (x - y)/\sqrt{2} ,$$
$$y' = (x + y)/\sqrt{2} . \tag{B.2}$$

Next, we "squeeze" both axes using the arctan coordinates to map the values of x' and y' from $(-\infty, +\infty)$ into a finite range $(-\pi/2, \pi/2)$; see Sect. 2.2.3 on arctan coordinates:

Fig. B.3 First step required
for Penrose transformations:
45° counter-clockwise
rotation of the axes

(a)

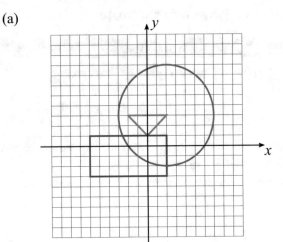

(b)

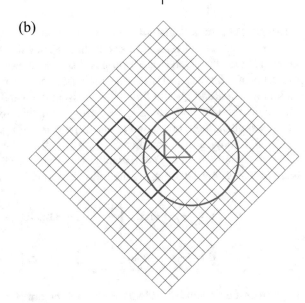

$$x'' = \arctan x' \, ,$$
$$y'' = \arctan y' \, . \tag{B.3}$$

Finally, we rotate the axes x'' and y'' by $\pi/4$ counter clock-wise:

$$X = (x'' + y'')/\sqrt{2},$$
$$Y = (x'' - y'')/\sqrt{2}. \tag{B.4}$$

Fig. B.4 Second and third
steps required for Penrose
transformations: (**a**)
"Compression" of an infinite
plane into a finite range; (**b**)
45° clockwise rotation

(a)

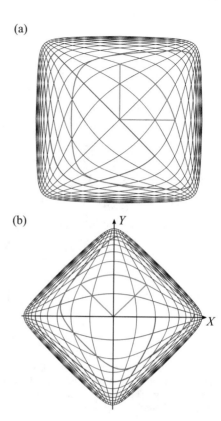

(b)

When applied to spacetime coordinates (t, x), Penrose transformation yields two
independent coordinates, T and X, that can be used to plot events and worldlines
just as well as using Minkowski diagrams with the axes x and t. The lines of constant
location x and the lines of constant time t are not straight lines in Penrose diagrams.
However, the worldlines of light remain diagonal, as in Minkowski diagrams; see
Fig. B.5.

Penrose diagrams show whole infinite spacetime at once: the events in the infinite
future and in the infinite past all can be seen in the diagram (events 1 and 3 in
Fig. B.5). Similarly, the events that happen infinite distance away from the origin
event (events 2 and 4) also fit into a finite drawing.

B.4   Generalized Lorentz Transformation

The Lorentz transformation

$$t' = \frac{t - vx}{\sqrt{1 - v^2}},$$

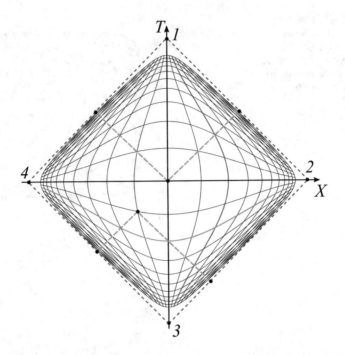

Fig. B.5 Infinite spacetime represented in Penrose coordinates. Similar to Minkowski diagrams, worldlines of light signals are diagonal. The world lines of stationary objects ($x = const$) are not straight; the lines of simultaneity ($t = const$) are also curved. The temporal (1) and (3) and spatial (2) and (4) infinities are all present in the diagram

$$x' = \frac{x - vt}{\sqrt{1 - v^2}}$$

was derived in this book using the light postulate and the radar method. Many other derivations also rely on the use of the light postulate. However, as has been known since 1910, the form of the relationship between spacetime coordinates (t', x') and (t, x) can be deduced from more basic principles. It the works of Ignatowsky,[4] Frank and Rothe[5] it has been shown that the relationship can be written as

$$t' = \frac{t - vx/\sigma^2}{\sqrt{1 - v^2/\sigma^2}},$$

$$x' = \frac{x - vt}{\sqrt{1 - v^2/\sigma^2}},$$

---

[4] Ignatowsky W A *Phys. Z.* **11** 972 (1910).

[5] Frank P, Rothe H *Ann. Phys.* **34** 825 (1911).

where σ is a quantity with the dimension of velocity. Importantly, any "particle" moving with the velocity σ in one inertial reference frame, will have the same velocity σ in other inertial reference frames. It is worth getting acquainted with general arguments leading to this result.

Consider two inertial observers, S and S', moving relative to each other with the velocity v. If S and S' observe a pair of freely floating (i.e., free from external forces) objects at rest relative to each other, they represent the objects' worldlines as parallel straight lines in their respective Minkowski diagrams. Therefore, the transformation between spacetime coordinates (t, x) and (t', x') of *the same events* can, in general, be written as follows:

$$x' = Ax + Bt,$$

$$t' = Cx + Dt.$$

Such transformation, called linear transformation, transforms a straight line in (t, x) into a straight line in $(t'x')$; furthermore, parallel lines in (t, x) remain parallel in (t', x'). Physically it means that objects moving with equal velocities u relative to S, move with equal velocities $u' \neq u$ relative to S'. The presence of terms quadratic or higher power of either t or x will transform a straight line $t = kx + b$ into a curved line in t' vs x'. The coefficients A, B, C, and D, are not constant numbers, they may depend on the relative velocity v:

$$A = A(v), \ B = B(v), \ C = C(v), \ D = D(v).$$

There must exist a relationship between the coefficients A and B. To see this, consider, for example, how the observer S describes the motion of the observer S': the events that correspond to the origin of the S' are all given by $x' = 0$, whereas relative to S these events all lie on the worldline $x = vt$, describing the uniform rectilinear motion of S' relative to S. Thus, from

$$Ax + Bt = x' = 0$$

follows

$$x = -\frac{B}{A}t = vt.$$

From the last expression we find

$$B(v) = -vA(v).$$

Using this relationship, we can write

$$x' = Ax - vAt = A(v)(x - vt).$$

Imagine that the observer S measures a distance $\Delta x > 0$ between a pair of simultaneous events ($\Delta t = 0$). If the axes x and x' point in the same direction (our default assumption), we get

$$\Delta x' = A(v)\Delta x > 0.$$

Furthermore, we expect $\Delta x'$ to be in the same relation to Δx (greater or smaller) regardless of the direction of relative motion of S' and S. In other words, by flipping the direction of motion we can not flip the locations of two events: the one on the right remains on the right. Thus

$$\Delta x' = A(v)\Delta x = A(-v)\Delta x,$$

which implies

$$A(-v) = A(v).$$

Since the observers S and S' are equivalent, *the form of the relationship* between (t', x') and (t, x) must be the same. Obviously, the velocity of S relative to S' is the opposite to the velocity of S' relative to S, but this is the only difference between the observers. Thus, we must have

$$x = A(-v)(x' + vt') = A(v)x' + vA(v)t'.$$

Now, using the expression for x' obtained above, we rewrite the last formula as follows

$$x = A^2(v)x - vA^2(v)t + vA(v)t',$$

from which we immediately find t' in terms of t and x:

$$vA(v)t' = vA^2(v)t + x[1 - A^2(v)].$$

It can be further simplified

$$t' = A(v)\left(t - \frac{A^2(v) - 1}{vA^2(v)}x\right).$$

To make it more similar to the formula for x', we will explicitly write t' as the function of vx:

$$t' = A(v)\left(t - \frac{A^2(v) - 1}{v^2A^2(v)}vx\right).$$

Thus, we got rid of all the coefficients B, C, D, rewriting everything in terms of an unknown function of velocity $A(v)$.

To summarize all we found so far, the relationship between spacetime coordinates of S and S' can be written as follows:

$$x' = A(v)(x - vt),$$

$$t' = A(v)(t - K(v)vx).$$

Where we introduced a helper function

$$K(v) = \frac{A^2(v) - 1}{v^2 A^2(v)}.$$

The "inverse" transformation is given by

$$x = A(v)(x' + vt'),$$

$$t = A(v)(t' + K(v)vx').$$

If we can find $A(v)$, then we can calculate $K(v)$ and complete the transformation. Also, if we can find $K(v)$, we can find $A(v)$ (up to the sign, since K is related to A^2).

Composition of Transformations

To determine the functions $A(v)$ and $K(v)$ we will consider three observers: S, S', and S''. The observer S' is moving relative to S with the velocity v_1, and the observer S'' is moving relative to S' with the velocity v_2; see Fig. B.6. We will first find the velocity of S'' relative to S, using the "direct" transformation formulas. Next, we will find the velocity of S relative to S'', using the "inverse" transformation formulas. The answer, obviously, must be the same, except for the flipped sign of relative velocity of S and S''.

To reduce clutter, we will use the following notation

$$A_1 = A(v_1) = A(-v_1), \ K_1 = K(v_1) = K(-v_1),$$

and similarly for other velocities.

First, the spacetime coordinates of the same event as measured by S'' and S' are related

$$x'' = A_2(x' - v_2 t'),$$

$$t'' = A_2(t' - K_2 v_2 x').$$

Using similar equation for x' and t', substituting them in the expressions above, we arrive at

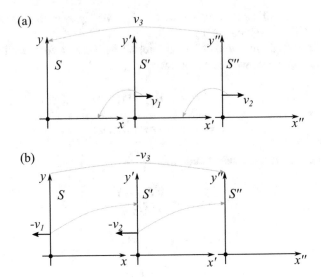

Fig. B.6 The relationship between the spacetime coordinates of the reference frames S and S'' can be calculated using "direct" transformation (**a**) and "inverse" transformation (**b**). The only difference in the results comes from the direction of the relative velocity $\mathbf{v}_3$

$$x'' = A_1 A_2([1 + K_1 v_1 v_2]x - [v_1 + v_2]t).$$

We can rewrite the last equation to look more like

$$x'' = A_3(x - v_3 t),$$

where v_3 is the velocity of S'' relative to S:

$$x'' = A_1 A_2 [1 + K_1 v_1 v_2] \left(x - \frac{v_1 + v_2}{1 + K_1 v_1 v_2} t \right).$$

From this formula we find that

$$v_3 = \frac{v_1 + v_2}{1 + K_1 v_1 v_2};$$

this is the *velocity composition formula* in general form.

Next, we will combine the "inverse" transformation to find $-v_3$. First, we write

$$x = A_1(x' + v_1 t').$$

Second, recall that

$$x' = A_2(x'' + v_2 t''), \quad t' = A_2(t'' + K_2 v_2 x'').$$

Substituting these two into the expression for x vs (t', x'), we arrive at

$$x = A_1 A_2 [1 + K_2 v_1 v_2] \left(x'' + \frac{v_1 + v_2}{1 + K_2 v_1 v_2} t'' \right).$$

Comparing this expression with the expected form

$$x = A_3 (x'' + v_3 t''),$$

we conclude that v_3 can be written in two ways:

$$v_3 = \frac{v_1 + v_2}{1 + K_1 v_1 v_2},$$

or

$$v_3 = \frac{v_1 + v_2}{1 + K_2 v_1 v_2}.$$

The important thing is that in the first expression there is a function $K(v_1)$, while in the second the functions is $K(v_2)$. From the equality of both forms of v_3 *for arbitrary velocities* v_1 and v_2, follows that

$$K(v_1) = K(v_2) = K = const.$$

Now we can find the function $A(v)$ in terms of this constant:

$$A(v) = \frac{1}{\sqrt{1 - K v^2}}.$$

The constant K must be a quantity with the units of the inverse velocity squared, so that $K v^2$ becomes dimensionless. Let's denote

$$K = \frac{1}{\sigma^2}, \quad \rightarrow \quad \sigma = \frac{1}{\sqrt{K}},$$

where σ is yet undetermined velocity. The generalized transformation of spacetime coordinates thus takes form

$$x' = \frac{x - vt}{\sqrt{1 - v^2/\sigma^2}},$$

$$t' = \frac{t - vx/\sigma^2}{\sqrt{1 - v^2/\sigma^2}}.$$

The inverse transformation is obtained by flipping the sign of the relative velocity v.

Invariant Velocity

Using the velocity composition formula, derived above:

$$v_3 = \frac{v_1 + v_2}{1 + K v_1 v_2} = \frac{v_1 + v_2}{1 + v_1 v_2 / \sigma^2},$$

we find that if $v_1 = \sigma$, then

$$v_3 = \frac{\sigma + v_2}{1 + v_2/\sigma} = \sigma;$$

similarly for $v_2 = \sigma$. Thus, the velocity σ is the same for all reference frames. It is some kind of invariant velocity, yet undetermined.

In conclusion, we found a generalized form of spacetime coordinate transformation without relying on the invariance of the speed of light (light postulate). We used very simple assumptions about the kind of transformation of spacetime coordinates are reasonable between two inertial observers; additionally, we used the equivalence of the inertial observers. We also discovered that there must exist some invariant velocity. The experiment must determine its physical meaning and value.

B.5   Lengths of Objects

The special theory of relativity presents some pre-relativistic concepts in a new light. One such important concept is the concept of an *object*, especially an *extended object*. Among the properties of an extended object, pre-relativistic worldview considers its *length* as an important, even an objective property. However, according to the special theory of relativity, neither the concept of an extended object, nor the notion of its length are absolute.

B.5.1   Objects

A galaxy, a star, a rock, a person, a biological cell, a molecule, an atom, an electron—these are objects of various degrees of complexity. In the modern view of the world, none of them are eternal: They form, exist for some time, and evolve into some new object or complex of objects. The special theory of relativity, with its emphasis on events and the relationships between them, leads to a different view on the objects. As Bertrand Russell, in *"ABC of Relativity"*, beautifully expressed it:

▷ **Russell on Objects**

These are really mere mathematical constructions out of events, but owing to their permanence they are practically important, and our senses (which were presumably developed by biological needs) are adapted for noticing them, rather than the crude continuum of events which is theoretically more fundamental.

Consider, for example, a relatively simple process, involving subatomic particle, such as a muon, a neutron, a proton, an electron, and an electron anti-neutrino; it is shown in Fig. B.7a. In this process, an energetic muon collides with a stationary atom and knocks out a neutron from its nucleus. The neutron eventually decays, producing the proton, the electron, and the electron anti-neutrino, as illustrated in Fig. B.7a.

Among the manifold of events, we highlight the worldline of a muon, because of its persistent properties, such as electric charge, spin, and mass, to name a few. The same applies to the worldlines of the atom, the neutron, and the rest of the particles: To each event on the worldline of a given particle we assign the same set of *physical properties*, and, as long as these properties persist, we consider the worldline as describing the same object—a subatomic particle, for example.

An extended macroscopic object is comprised of a myriad of molecules, atoms, or subatomic particles. A piece of a rock with the size of a thumbnail contains about 10^{23} of atoms—almost ten orderx of magnitude more than the number of

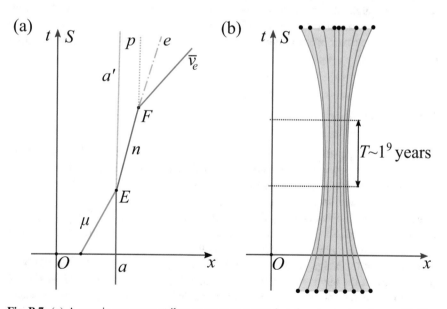

Fig. B.7 (a) A cosmic-ray muon strikes an atom at rest to knock out a neutron (event E). The latter decays into a proton, an electron, and an electron anti-neutrino (event F). (b) Minkowski diagram illustrating the evolution of an average star: A gas cloud (bottom dots) collapses, creating a dense core of a star; the star then emits light for several billion years (worldlines of light not shown). Eventually, the star may shed the outer layers and its core may become even denser. The details of the evolution depend on the mass of the star, but are not very important here

stars in a giant galaxy. The Minkowski diagram of such a piece of rock, if plotted in corresponding detail, would contain a mind-boggling number of worldlines, each representing a separate atom. Let us examine, instead, the simplified Minkowski diagram of an average star, as illustrated in Fig. B.7b.

Stars develop from collapsing clouds of gas and dust. As a cloud becomes denser, the emerging "core" becomes hotter, and, eventually, ignites a chain of nuclear fusion reactions, that keep the star emitting light. After several billion years, the "core" of the star becomes denser, and, under certain conditions, the star can shed some of its outer layers, as schematically depicted in Fig. B.7b.

The details of this particular stellar evolution are not important. What is important is the *notion of the star as an object*. Astrophysicists assign a mass, an angular momentum, a temperature, and a size or shape to a star. Those parameters are not constant, they depend on the age of the star, which is an additional parameter describing the star as a unique astrophysical object. Thinking about the star as a single object, with certain physical properties having well-defined values at a given moment of time, we choose a specific reference frame (e.g., the one where the core of the star is at rest), and consider all events, happening in different places, *at the same time*. However, according to the special theory of relativity, *there is no absolute/invariant notion of an extended object, nor of a property of such an object that relies on the idea of simultaneity (e.g., length)*. For example, the diameter of the star, although an important parameter, is essentially a length-like quantity, and, therefore, depends on the frame of reference.

It should be clear: What applies to a star, applies to any other spatially extended object.

B.5.2   Lengths

Measurements are affected by various conditions: by the relative location of an observer and an object being measured; by the interaction between an observer (or a measurement apparatus) and an object being measured; and by the relative motion of an observer and an object being measured. An example of the first kind is the *apparent size* of the object, when the object becomes smaller as the distance between the observer and the object increases. An example of the second kind is the *measured temperature* of the object using a thermometer with comparable heat capacitance, when the heat exchange between the thermometer and the object changes the final temperature of the latter. An example of the third kind is the size of the object in the direction of its motion, when the *measured length* is becoming shorter as the speed of the object relative to the observer increases. In each case we can inquire *"Why does this happen?"* or *"What causes this effect?"* The answers, depending on the case, will be quite different.

The farther an object is located relative to an observer, the smaller it appears. It happens because the propagation of light is determined by the geometry of space, and because the latter is Euclidean.[6] There appears to be nothing that *causes* this effect, in the sense in which cooling an object causes it to shrink. It is a *geometrical effect*, affecting all visible dimensions of all bodies in the same way.

When the temperature of an object is measured by connecting it to a thermometer, the two exchange thermal energy until their temperatures equalize. If the initial temperature of the thermometer is higher than that of the object, the object will gain some thermal energy, ending up with higher temperature than before the measurement. The opposite will happen if the thermometer is initially colder than the object. This happens because for objects in thermal contact the energy flows from places of higher temperature into places of lower temperature and losing or gaining energy usually (but not always) requires a temperature change. At the microscopic level, the chaotic motions and collisions (*interactions!*) of particles, making up the bodies, cause the energy flow from hot regions to cold ones.

When measuring the length of an object we can locate two points, say the front and the rear, and measure the distance between them. For a moving object the selected points must be located at the same time, if we want to use this straightforward method, otherwise the point measured second will move, making the measured length either shorter or longer. The result of such measurement is counter-intuitive: the length, measured by simultaneously locating two points, will depend on the relative speed of the object and the observer. Moreover, the faster the object is moving, the smaller will be its measured length; this is true for all objects. The described effect is a *kinematic effect*—an effect of relative motion, in contrast to *dynamic effects* due to physical interactions of objects. It is more akin to the geometric effect of apparent size change due to relative distance, than to the effect of the temperature change when contacting with a thermometer. There is nothing that *causes* (in the sense in which external stresses cause bodies to shrink) all lengths to change with velocity.

If the lengths of all moving objects are affected by their motion relative to an observer, should we conclude that it is the "space itself", which contains all object, is affected? This line of reasoning is unhelpful. Consider, for example, two identical meter sticks traveling with different velocities relative to a given observer. Their measured lengths will be different, while their lengths at rest (proper lengths) are the same. If the space itself is affected, should it not affect all things contained in it in the same way? Or should we admit that objects moving with different velocities, including objects at rest, have their own spaces? Besides, the effect of relative motion on lengths has more to do with time measurements, than with space-related measurements, as the special theory of relativity reveals.

One final analogy might be helpful. As shown in Fig. B.8, for the same planar geometric figure (square in this case), its "shadow" on the x axis will have different length for different orientations of the figure relative to the "rays of light" or the axes. All lengths are equally valid and can be useful in solving problems involving

---

[6] In non-Euclidean geometry the apparent size of an object may increase with the increasing distance.

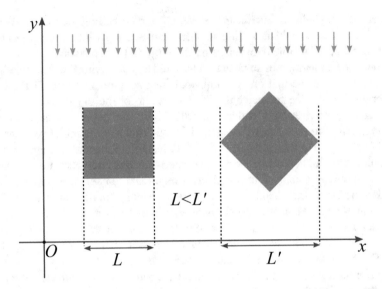

Fig. B.8 For different orientations of the square relative to the rays of light, the length of the shadow will be different. None of the possible lengths is the *true* length

the figure. In the special theory of relativity, a region of spacetime corresponds to a figure, and its one specific *simultaneous slice* corresponds to the determination of the length of an "object". None of the possible lengths is the *true* length; all lengths are equally valid and can be useful in solving problems involving the events making up the region of spacetime.

B.6   Additional Problems

B.6.1   Geometry

? Problem B.1 *

Consider an oblique coordinate system, which is like Cartesian, except that the angle between the axes is not 90°. Figure B.9 shows a special case of oblique coordinates, where the axes y' and x' are symmetrically rotated towards the diagonal line. Find the transformation of coordinates between such coordinate system and Cartesian coordinates, assuming that they both share the same origin, as shown in Fig. B.9.

Fig. B.9 Transformation of
the coordinates of a point P
from a Cartesian coordinate
system S into an oblique
coordinate system S'

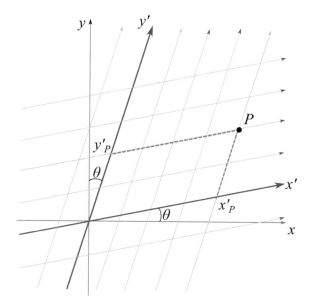

Consider three Cartesian coordinate systems, shown in Fig. B.10. The rotation of
C' relative to C by an angle α can be followed by the rotation of C'' relative to C'
by an angle β, resulting in the rotation of C'' relative to C by an angle $\phi = \alpha + \beta$.
Use this fact, in combination with the transformation equations for the Cartesian
coordinates, to show that

$$\cos(\alpha + \beta) = \cos\alpha\cos\beta - \sin\alpha\sin\beta,$$

and

$$\sin(\alpha + \beta) = \sin\alpha\cos\beta + \sin\beta\cos\alpha.$$

B.6.2   Galilean Relativity

A group of joggers is running from left to right, forming a line of length L, as shown
in Fig. B.11. Each runner has the same speed v. As soon as a runner reaches a coach

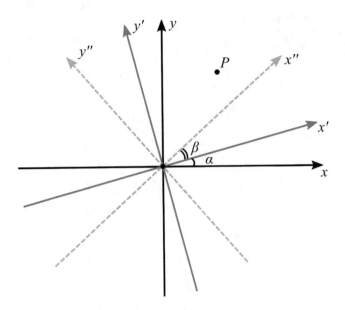

Fig. B.10 Cartesian coordinates C' are rotated relative to the Cartesian coordinates C by an angle α. Cartesian coordinates C'' are rotated relative to the Cartesian coordinates C' by an angle β. Relative to the Cartesian coordinates C, the axes of C'' are rotated by $\phi = \alpha + \beta$

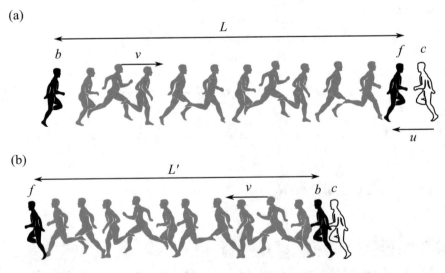

Fig. B.11 A group of joggers runs towards the coach and then turns around. What is the final distance between the front and the back of the line? For Exercise B.3

(shown as an outlined figure with c on top), she turns around and keeps running in the opposite direction with the same speed v.

Find the distance L' from the front to the back runner after all runners met the coach. The coach is moving from right to left with the speed $u < v$.

B.6.3 Special Relativity

? Problem B.4

Suppose that instead of the spacetime coordinates (t, x) an observer A uses transformed coordinates

$$\eta = t - x, \qquad \tau = t + x.$$

They completely and uniquely specify an event, and therefore are just as good as t and x. In Mikowski diagrams the axes η and τ will be rotated relative to t and x by 45°, and go diagonally, parallel to the light-lines.

Consider an observer B, that also uses transformed coordinates

$$\eta' = t' - x', \qquad \tau' = t' + x'.$$

Find the relationship between the transformed coordinates (η', τ') and (η, τ).

? Problem B.5

A rigid rod is approaching an absolutely rigid wall with speed v (see Fig. B.12). Given the proper length of the rod L_0, find the minimum length of the rod during the collision. Draw Minkowski diagram of the collision process.

Fig. B.12 A rigid rod is colliding with an absolutely rigid wall and is compressed in the process. What is the minimum length during the collision? For Exercise B.5

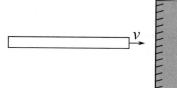

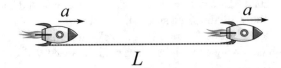

Fig. B.13 Two identical spaceships, connected with an unstretchable string, start their engines simultaneously. Will the string remain unbroken? For Exercise B.6

? Problem B.6

Two identical spaceships are at rest in some reference frame. The distance between the ships is L. The spaceships are connected by an unstretchable string with the proper length L (see Fig. B.13). At some moment, both spaceships start their engines (*at the same time!*) and start moving with identical accelerations. Will the string remain unbroken as the ships keep moving?

? Problem B.7

Laser Interferometer Gravitational-Wave Observatory (LIGO) has two laboratories, separated by a distance 3,002,000 meters. In the morning of September 14, 2015, both laboratories detected a gravitational wave signal from a pair of black holes merging into one. Assuming that the signal was received by both laboratories simultaneously, find the time interval between the detections of the signal from the point of view of an observer at rest relative to the Sun. The orbital speed of the Earth is 30,000 meters per second.

? Problem B.8

Observers A and B are moving relative to each other with speed $v = 0.6$. When they meet, they synchronize their clocks. Later, when the clock of A indicates time $t = 1\text{s}$, A emits a light-pulse towards B. The observer B reflects this pulse back to A, which reflects it back to B and so on; the light-pulse is bouncing between A and B.

(A) Draw Minkowski diagram of this scenario.
(B) Find the time of the first 3 reflection events, as measured by each observer.

? Problem B.9

Observer A measures three events, E, F, and G, as happening at the following spacetime coordinates: E at $(0s, 2s)$; F at $(1s, 4s)$; and G at $(2s, 6s)$.

(A) What should be the velocity of an observer B, so that B measures these events as simultaneous.
(B) Assume that B has the speed $v = 0.9$. Find spacetime coordinates of the events E, F, and G, as measured by B. What is their order?
(C) Draw Minkowski diagram for the reference frame of A. Plot the world line of B. Indicate events E, F, and G. In what order will the observer B see (that is, will receive light signals from) the events E, F, and G.

B.6.4 Energy-Momentum

? Problem B.10

A photon and an electron are moving towards each other, as shown in Fig. B.14. The energy of the photon is E. After the collision the electron stops, while the photon is scattered in the direction opposite to the original direction of motion. Find the energy of the scattered photon.

Fig. B.14 A photon gains energy by scattering from a fast moving electron. The electron is stopped as the result. For Exercise B.10

(a)

(b)

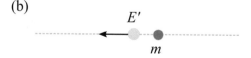

Answers

B.7 Geometry

Problem 2.1
NA.

Problem 2.2
In the cases (A) and (B) the area is the same—4. It is the area of a square with the side 2. In the cases (C), the area is 1; it is the area of a square with the side 1.

Problem 2.3

$$r = +\sqrt{x^2 + y^2},$$
$$\phi = \arctan(y/x)$$

Problem 2.4

$$x = x' \cos\theta + y' \sin\theta,$$
$$y = x' \sin\theta + y' \cos\theta.$$

Problem 2.5
NA.

Problem 2.6
Surfaces are described by the equation

$$z = \frac{y^2}{4x}$$

© The Author(s), under exclusive license to Springer Nature Switzerland AG 2022
Y. Deshko, *Special Relativity*, Undergraduate Lecture Notes in Physics,
https://doi.org/10.1007/978-3-030-91142-3

Problem 2.7
Going from the state 1 to 2 the gas is getting colder. Going from the state 2 to 3 the gas is getting hotter and expands. Going from the state 3 to 1 the temperature can not remain constant during this process.

$$T_1 = \frac{P_1 V_1}{Nk},$$

$$T_2 = \frac{P_2 V_2}{Nk},$$

$$T_3 = \frac{P_3 V_3}{Nk}.$$

B.8   Galilean Relativity

Problem 4.1

$$v + 2u.$$

Problem 4.2

$$\phi = \tan^{-1} \frac{u}{v}.$$

Problem 4.3

$$\sin \alpha = \frac{u}{v}.$$

Problem 4.4

$$d = D \frac{v \cos \phi + u \cos \theta}{\sqrt{v^2 + u^2 - 2uv \cos(\theta - \phi)}}.$$

Problem 4.5

$$L = uT = 2D \frac{v}{u}.$$

B.9 Spacetime

Problem 3.1
NA.

Problem 3.2

$$t_1 = t - |x|, \quad t_2 = t + |x|.$$

Problem 3.3
NA.

Problem 3.4
NA.

Problem 3.5
The observer A will first "see" the clock of the observer B ticking faster than its own. Then, after the observers meet, the clock of B will appear to be ticking slower, compared to the clock of A.

B.10 Special Relativity

Problem 5.1
Acceleration has the units of inverse seconds, or Hertz, in c-based units.

Problem 5.2
NA.

Problem 5.3
NA.

Extended Problem: Relativity of Simultaneity

NA.

Problem 5.4

$$T_R^{(B)} = k^{(B)} T_E^{(B)},$$

where

$$k^{(B)} = 1 + v.$$

Problem 5.5

$$k = \sqrt{\frac{1+v}{1-v}}.$$

Problem 5.6

$$v = \frac{k^2 - 1}{k^2 + 1}.$$

Problem 5.7

$$z = k - 1 = v.$$

Problem 5.8
When two observers are approaching each other, the time intervals between emitted and received signals are related as

$$T_R = \frac{T_E}{k},$$

where k is Doppler factor for the original scenario of receding observers.

Problem 5.9

$$v = \frac{(z+1)^2 - 1}{(z+1)^2 + 1}.$$

For $z = 0.01$ the relative speed is $v = 0.01$; for $z = 0.1$, $v = 0.095$; for $z = 1.0$, $v = 0.6$; for $z = 10.0$, $v = 0.98$.

Problem 5.10

$$\bar{v} = 22 \times 10^3 \text{ km/s}.$$

Problem 5.11

$$v = \frac{1}{\sqrt{1 + (T_{ship}/d)^2}}.$$

Numerically, $v = 0.9994$.

Problem 5.12

$$\frac{t_R}{T'} = \gamma > 1.$$

Problem 5.13

$$w = \frac{(1+v)^N - (1-v)^N}{(1+v)^N + (1-v)^N}.$$

Problem 5.14

$$\gamma = \frac{1}{2}\left(k + \frac{1}{k}\right).$$

Problem 5.15

$$t = \gamma T.$$

Problem 5.16
NA.

Problem 5.17

$$t = \gamma(t' + vx'),$$

$$x = \gamma(x' + vt').$$

Problem 5.18

$$\bar{t}' = \frac{\bar{t} - \bar{v}\bar{x}/\bar{c}^2}{\sqrt{1 - \bar{v}^2/\bar{c}^2}}.$$

$$\bar{x}' = \frac{\bar{x} - \bar{v}\bar{t}}{\sqrt{1 - \bar{v}^2/\bar{c}^2}}.$$

Problem 6.1
If $u_c \to \infty$, the required speed becomes infinitely small.

Problem 6.2

$$\Delta t' = -2573 \text{ years.}$$

Problem 6.3

$$L_1 = 0.95L_0, \quad L_2 = 0.6L_0.$$

Problem 6.4
NA.

Problem 6.5

$$v = \frac{D}{\sqrt{D^2 + T_0^2}}.$$

Problem 6.6
Straight line is the longest. Zig-zagging light-lines—the shortest.

Problem 6.7
NA.

Problem 6.8
If the velocity v satisfies the condition

$$v = \frac{\Delta t_{12}}{\Delta x_{12}} = v_0 < 1,$$

then the events will be simultaneous for B.

Problem 6.9

$$t' + vx' = 0, \quad \rightarrow \quad t' = -vx'.$$

Problem 7.1

$$t = \gamma(t' + vx'),$$
$$x = \gamma(x' + vt'),$$
$$y = y',$$
$$z = z'.$$

Problem 7.2

$$u'_x = \frac{u_x - v}{1 - vu_x},$$

$$u'_y = \frac{u_y}{\gamma(1 - vu_x)},$$

$$u'_z = \frac{u_z}{\gamma(1 - vu_x)}.$$

Problem 7.3
NA.

Problem 7.4
NA.

Problem 7.5
NA.

Problem 7.7

$$p'_t = \gamma m, \quad p'_x = -\gamma mv.$$

Problem 7.8

$$M' = \sqrt{M(M + 2E)}.$$

Problem 7.9

$$v = E/(M - E).$$

Problem 7.10

$$v = \frac{\sqrt{M(M + 4m)}}{M + 2m}$$

or

$$v = \sqrt{1 - 1/(1 + M/2m)^2}.$$

Problem 7.11

$$P = M/2.$$

Problem 7.12

$$E_k = 2(M - m).$$

Problem 7.13

$$v = \cos \frac{\alpha}{2}.$$

Problem 7.14

$$v = \frac{E}{\sqrt{E^2 + M^2}}.$$

B.11   Additional Problems

Problem B.9

$$x' = \frac{x \cos \theta - y \sin \theta}{\cos 2\theta},$$
$$y' = \frac{-x \sin \theta + y \cos \theta}{\cos 2\theta}.$$

Problem B.2
NA.

Problem B.3

$$L' = L \frac{v - u}{u + v}.$$

Problem B.4

$$\eta' = k\eta,$$

where

$$k = \sqrt{\frac{1 + v}{1 - v}} = \gamma(1 + v)$$

is the Doppler k-factor. Similarly for τ':

$$\tau' = \tau/k.$$

Problem B.5

$$L_{min} = L_0\sqrt{\frac{1-v}{1+v}}.$$

Problem B.6
The proper length of the string must grow and the string has to snap.

Problem B.7

$$\Delta t' \approx 3 \times 10^{-6}\,\text{s}.$$

Problem B.8
The signal emitted at time T by the observer A will be received and reflected by the observer B at $2T$ (as measured by B). The reflected signal will be received back by A at time $4T$ (as measured by A). The signal reflected by A will be finally received by B at time $8T$.

According to A, the first reflection of the signal by B happens at

$$t_E = \frac{5}{2}T.$$

The observer B measures the event F at

$$t_F = 5T.$$

Problem B.9

(A) For $v = 0.5$, all three events, E, F, and G will be measured by the observer B to happen at the same time.
(B) The events will get the following spacetime coordinates in B's reference frame:

$$E \to (-4.1, 4.6), \quad F \to (-6.0, 7.1), \quad G \to (-7.8, 9.6).$$

(C) NA.

Problem B.10

$$E' = \frac{E}{1 - 2E/m}.$$

Hints

B.12 Geometry

Problem 2.1
Consider two right triangles, $\triangle ACD$ and $\triangle BED$; use the fact that the sum of angles in any triangle equals π.

Problem 2.2
(A) Rewrite the inequality for the absolute value as two separate inequalities, then identify all points that satisfy the inequalities. The points will make a simple geometric figure.

(B) Same as in (A). In addition, identify where the lines, resulting from inequalities, intersect the coordinate axes.

(C) Same as in (B). In addition, identify where the lines, resulting from different inequalities intersect each other; that is, find he vertices of the figure.

Problem 2.3
For r—use Pythagoras' theorem. For the polar angle, ϕ, analyze different cases ($x = 0$, $y \neq 0$, etc.) and use arctan function to find the angle in terms of x and y.

Problem 2.4
Multiply the transformation equations by suitable factors to make some terms on the right cancel each other, if the left-hand and right-hand sides of two equations are added. Then do the same, to make some terms on the right cancel each other, if the equations are subtracted.

Problem 2.5
Add the squared differences of the primed coordinates, and cancel terms with the product of $\sin \theta$ and $\cos \theta$.

© The Author(s), under exclusive license to Springer Nature Switzerland AG 2022
Y. Deshko, *Special Relativity*, Undergraduate Lecture Notes in Physics,
https://doi.org/10.1007/978-3-030-91142-3

Problem 2.6

From the requirement of zero discriminant obtain the equation of the surface in the form $z = f(x, y)$. Then analyze the slices of this surfaces for several fixed values of x, then for several fixed values of y.

Problem 2.7

Observe which parameters remain the same in the process $2 \rightarrow 3$, and in $3 \rightarrow 1$. Use the ideal gas equation to find the behavior of the other parameters, as well as the temperature in the states 1, 2, and 3.

B.13 Galilean Relativity

Problem 4.1

Consider the problem in the reference frame of the wall.

Problem 4.2

Consider the problem in the reference frame of the cart.

Problem 4.3

Starting from the reference frame of the unstable particle, switch into the laboratory reference frame and use vector addition of velocities to determine all possible direction of the product particles. Note which direction results in the maximum angle of the combined velocities of the original and product particles.

Problem 4.4

Consider the problem in the reference frame of one of the cars. The direction of the trajectory of the other car will be determined by the components of its velocity relative to the first car; the components for each axis can be found using addition of velocities formula.

Problem 4.5

Remember that the cockroach is running with constant speed for a fixed time interval.

B.14 Spacetime

Problem 3.1

Use some common source of a "trigger" signal, placed in such a way, as to arrive to both clocks at the same time.

Problem 3.2

Try adding and then subtracting the expressions for t and x.

Problem 3.3
Consider two events in different places, making sure they lie on the same line of $t = const$.

Problem 3.4
Draw spacetime diagrams and observe that, as the distance between the objects changes, each successive signal takes different time to reach the target.

Problem 3.5
Use the result of the Problem 3.4.

B.15   Special Relativity

Problem 5.1
Use the definition of the acceleration in the meter-based units, and express the velocity and time in c-based units.

Problem 5.2
Firstly, use the fact that the speed of the observer A relative to B is the same as the speed of B relative to A. Secondly, remember that the worldlines of light are diagonal lines, going symmetrically between the spatial and temporal axes for all observers.

Problem 5.3
Look up the idea of "quantum vacuum".

Extended Problem: Relativity of Simultaneity
NA.

Problem 5.4
Consider a pair of signals emitted by A at two moments of time, when the distances between the observers, as measured by B, are D_1 and $D_2 > D_1$. Determine when B receives the emitted signals, taking into account their finite propagation speed.

Problem 5.5
Add one more signal, assuming that the observer B is locating A at the event 4. Use the relations between the intervals of the emitted and received pairs of signals, as measured by the receding observers.

Problem 5.6
Start with the expression for k^2 and solve for v.

Problem 5.7
Recall that for small velocity $k \approx 1 + v$.

Problem 5.8
When "time direction" is reversed, the roles of the observers are reversed, but the events, and time intervals, as measured by clocks, remain the same.

Problem 5.9
Use the definition of z to find k, then use the result of the Problem 5.6.

Problem 5.10
Use the ratio of measured wavelengths to find k, then use the result of the Problem 5.6.

Problem 5.11
First, find the relationship between the events of departure and the arrival of the spaceship in the reference frames of the Earth and the spaceship. Then write the travel time in the Earth's reference frame, given the distance to Andromeda and the speed of the spaceship. Finally, solve for v.

Problem 5.12
Imagine the radar method is used by the first laboratory to locate the event when the muon and the second laboratory meet.

Problem 5.13
Recall that Doppler k-factors of observers are multiplied.

Problem 5.14
Starting with expression for Doppler k-factor in terms of v, first multiply it by

$$\sqrt{\frac{1+v}{1+v}},$$

and then by

$$\sqrt{\frac{1-v}{1-v}}.$$

to find two different relationship between k, γ, and v. Then get rid of v, to obtain γ in terms of k only.

Problem 5.15
Note that the meeting of B and C happens half-way between the departure of B and the arrival of C.

Problem 5.16

(a) Find the Doppler k'-factor for the observers A and C. Then add a light signal, emitted by A at the event 4, towards B. Use the relationships between the time intervals as measured by the receding observers A-B (Doppler factor k), and C-B (Doppler factor k').
(b) Redraw the spacetime diagram from the part (a) in the reference frame C, and use the expression for the Doppler k' factor for the receding B and C.

Problem 5.17

Expand the parentheses in the Lorentz transformation equations. Then multiply them by suitable factors, so that adding their left-hand sides and right-hand sides results in the cancellation of some terms.

Problem 5.18

Simply replace x with $\bar{x}/\bar{c}$, v with $\bar{v}/\bar{c}$, and t with $\bar{t}$ everywhere.

Problem 6.1

Note that the minimum speed of the observer B, required to flip the order of any causally connected events, becomes zero.

Problem 6.2

Use the Lorentz transformation equation for $\Delta t'$, and the fact that $\Delta t = 0$.

Problem 6.3

Remember that the measured lengths (the simultaneous distance between the front and the back) of the rulers will be different in the reference frame S.

Problem 6.4

Square both sides of the two Lorentz transformation equations, and then add their left-hand sides, and right-hand sides. Some terms will cancel, to give the desired result.

Problem 6.5

Write the interval squared between birth and decay of a muon in the reference frames of the muon and the laboratory. Use the relationship between the life-time, speed, and the distance traveled in the laboratory. Then solve for v.

Problem 6.6

(a) Use the path-dependence of time, and the fact that the straight worldline between a pair of events has the longest "distance" (proper time).
(b) Recall that light-like intervals have zero length.

Problem 6.7

First plot the equation $t^2 = 1 + x^2$: two curves, one for the positive t, the other for the negative t. Notice that for $x \gg 1$, we have $t^2 \approx x^2$—the lines approach the diagonal light-lines.

The curves for the equation $x^2 = 1 + t^2$ look like the curves for $t^2 = 1 + x^2$, once we "flip" the axes $x \to t$ and $t \to x$.

The curves for $t^2 = x^2$ are diagonal lines.

Problem 6.8

Use the Lorentz transformation for the time interval $\Delta t'_{12}$ and require $\Delta t'_{12} = 0$. Find the condition for the relative velocity, and show that it must be less than unity.

Problem 6.9

Use the inverse Lorentz transformation for time t to find all events that have $t = 0$. These events are simultaneous according to A.

Problem 7.1
Use the result of the Problem 5.17.

Problem 7.2
Start with the full Lorentz transformation for the change in spacetime coordinates Δt, Δx, Δy, and Δz. Then use the definition of the components of the coordinate velocity. Simplify to get rid of Δt.

Problem 7.3
Write the magnitude squared of the velocity in terms of u_x and u_y. Then simplify the expression using the following relationships:

$$\sin^2 \alpha = 1 - \cos^2 \alpha$$

and

$$\gamma^2 - 1 = \gamma^2 v^2.$$

You will have to re-arrange some terms to show that the denominator equals the numerator, and the magnitude of the velocity is 1.

Problem 7.4
Given that

$$\tan \theta' = \frac{\sin \theta'}{\cos \theta'},$$

and

$$\gamma \approx 1,$$

use the values for the $\sin \theta'$ and $\cos \theta'$ for $\theta = 90°$.

Problem 7.5
Substitute the values of the components u_t and u_x into the Lorentz transformation for the components of spacetime velocity.

Problem 7.6
Use the relationships between the components of spacetime velocities and "normal" velocities:

$$u'_x = \gamma(u')u', \quad u_x = \gamma(u)u, \quad u_t = \gamma(u).$$

Denote $\gamma(u')u' = w$, solve for w^2 in terms of v^2 and u^2. Then solve for u'.

Problem 7.7
Substitute the values of the components p_t and p_x into the Lorentz transformation for the components of spacetime momentum.

Problem 7.8

Use the conservation laws to find v. Then find $\gamma(v)$ to get

$$M' = \frac{E'}{\gamma}.$$

Problem 7.9

Use energy and momentum conservation. For speed, use relation between energy and momentum. See also Problem 7.8.

Problem 7.10

Use energy conservation and the expression for the relativistic energy to find γ. From γ find v.

Problem 7.11

Use the energy conservation and the relation between momentum and energy for massless particles.

Problem 7.12

Use the energy conservation and the definition of kinetic energy. The energy required to produce two protons is minimal when both protons are at rest.

Problem 7.13

Use the conservation of energy and the conservation of momentum for directions parallel and perpendicular to the original trajectory of the particle.

Problem 7.14

Use the conservation of energy and momentum and the relation between momentum, energy and speed.

B.16   Additional Problems

Problem B.9

First find the Cartesian coordinates x and y in terms of oblique coordinates. Then find the inverse transformation.

Problem B.2

Start with the expression of x'' in terms of x' and y', then plug in the expressions for x' and y' in terms of x and y. Then compare with the expected expression for x'' in terms of x and y and the total angle $\alpha + \beta$.

Problem B.3

Use the reference frame of the coach.

Problem B.4

Substitute Lorentz transformation into the definition of η' and τ' and simplify.

Problem B.5
The rod can not stop all at once. A finite time is required for the signal from the wall to reach, through the bulk of the rod, the moving left edge.

Problem B.6
Recall the relationship between the measured and the proper length of a moving object.

Problem B.7
Write the Lorentz transformation for the time $\Delta t'$ between the detection events. Then use the fact that they are simultaneous ($\Delta t = 0$) in the reference frame of the laboratories. Do not forget to consistently use correct units: either c-based or meter-based.

Problem B.8
Use the Doppler effect of "stretching" the time between the emitted and received signals, as they bounce between a pair of receding observers. Also, use the relationship between the emitted/received times and the time of reflection from the radar method.

Problem B.9

(A) Use the Lorentz transformation for the time interval between events to find the relative velocity v, required to make a pair of events $E - F$ simultaneous. Do the same for $F - G$.
(B) Use Lorentz transformation to find spacetime coordinates in the reference frame B.
(C) NA.

Problem B.10
Use the conservation of energy and momentum and solve for the energy after the scattering E'.

Solutions

B.17   Geometry

Problem 2.1

Given that the sum of angles in any triangle equals π, we may write for two right triangles, $\triangle ACD$ and $\triangle BED$ in Fig. 2.1:

$$\alpha + \gamma + \pi/2 = \pi = \beta + \gamma + \pi/2,$$

from this immediately follows $\alpha = \beta$.

Problem 2.2

(A) The condition $|x| \leq 1$ is equivalent to two independent conditions

$$x \leq 1,$$
$$x \geq -1.$$

The first of these inequalities describes all points in a plane on the left of the vertical line $x = 1$; the second inequality—all points on the right of the vertical line $x = -1$. The condition $|x| \leq 1$ describes all points in the vertical strip limited by the -1 and 1 on the x axis.

Similarly, the condition $|y| \leq 1$ describes all points in the horizontal strip limited by the -1 and 1 on y axis. Taken together, the conditions

$$|x| \leq 1,$$
$$|y| \leq 1,$$

describe the points inside the square centered at $(0, 0)$ and having the length of the side 2. The area of this figure is 4.

© The Author(s), under exclusive license to Springer Nature Switzerland AG 2022
Y. Deshko, *Special Relativity*, Undergraduate Lecture Notes in Physics,
https://doi.org/10.1007/978-3-030-91142-3

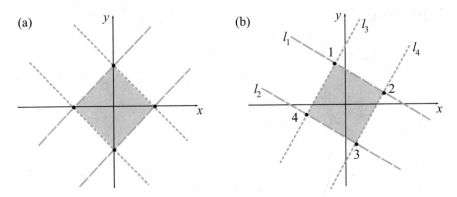

Fig. B.15 A figure in a plane can be specified by a set of inequalities for the coordinates x and y. The exact form of inequalities depends on the relative orientation of the figure and the Cartesian axes. In this problem a square with a side length 2 is used

(B) The condition $|x + y| \le \sqrt{2}$ is equivalent to two independent conditions

$$x + y \le \sqrt{2},$$
$$x + y \ge -\sqrt{2}.$$

The first of these inequalities describes all points in a plane below the line $y = \sqrt{2} - x$; the second inequality—all points above the line $y = -\sqrt{2} - x$. Both these lines have the negative slope of 45°, as shown in Fig. B.15a. The first line, $y = \sqrt{2} - x$, intersects the x axis at $\sqrt{2}$, and the y axis at $\sqrt{2}$. The second line, $y = -\sqrt{2} - x$, intersects the x axis at $-\sqrt{2}$, and the y axis at $-\sqrt{2}$.

Similarly, the condition $|x - y| \le \sqrt{2}$ is equivalent to two independent conditions

$$x - y \le \sqrt{2},$$
$$x - y \ge -\sqrt{2}.$$

The first of these inequalities describes all points in a plane above the line $y = -\sqrt{2} + x$; the second inequality—all points below the line $y = \sqrt{2} + x$. Both these lines have the positive slope of 45°. The first line, $y = -\sqrt{2} + x$, intersects the x axis at $\sqrt{2}$, and the y axis at $-\sqrt{2}$. The second line, $y = -\sqrt{2} - x$, intersects the x axis at $-\sqrt{2}$, and the y axis at $\sqrt{2}$.

The resulting figure looks like a square rotated 45°, with the vertices on the x and y axes. The length of the side is easily found to be 2. The area is 4, same as in the previous case.

(C) The condition $|x + \sqrt{3}y| \le 1$ is equivalent to two independent conditions

$$x + y\sqrt{3} \le 1,$$

$$x + y\sqrt{3} \geq -1.$$

The first of these inequalities describes all points in a plane below the line $y = 1/\sqrt{3} - x/\sqrt{3}$ (denote this line as l_1); the second inequality—all points above the line $y = -1/\sqrt{3} - x/\sqrt{3}$ (denote this line as l_2). Both these lines have the negative slope of 30°, as shown in Fig. B.15b.

Similarly, the condition $|y - \sqrt{3}x| \leq 1$ is equivalent to two independent conditions

$$y - \sqrt{3}x \leq 1,$$
$$y - \sqrt{3}x \geq -1.$$

The first of these inequalities describes all points in a plane below the line $y = 1 + x\sqrt{3}$ (denote this line l_3); the second inequality—all points above the line $y = -1 + x\sqrt{3}$ (denote this line l_4). Both these lines have the positive slope of 60°.

The first pair of parallel lines, l_1 and l_2, are perpendicular to the second pair of parallel lines, l_3 and l_4. The resulting figure limited by all four lines is a rectangle. To find the length of each side we can first find the Cartesian coordinates of the vertices and then use the distance formula between two points in a plane.

The lines l_1 and l_3 intersect at point (1), specified by the condition

$$\frac{1}{\sqrt{3}} - \frac{x}{\sqrt{3}} = 1 + x\sqrt{3},$$

from which we find the x coordinate of the intersection point (1):

$$x_1 = \frac{1 - \sqrt{3}}{4},$$

and corresponding y-coordinate is

$$y_1 = 1 + x_1\sqrt{3} = \frac{1 + \sqrt{3}}{4}.$$

The lines l_1 and l_4 intersect at the point (2), see Fig. B.15b s. The coordinates of this point can be found using the same approach as for the point (1). We first write the intersection condition

$$\frac{1}{\sqrt{3}} - \frac{x}{\sqrt{3}} = -1 + \sqrt{3},$$

from which follows

$$x_2 = \frac{1 + \sqrt{3}}{4},$$

and

$$y_2 = -1 + x_2\sqrt{3} = \frac{\sqrt{3}-1}{4}.$$

The coordinate distances between the points (1) and (2) along each axis are

$$\Delta x = \frac{1+\sqrt{3}}{4} - \frac{1-\sqrt{3}}{4} = \frac{\sqrt{3}}{2},$$

$$\Delta y = \frac{1+\sqrt{3}}{4} - \frac{\sqrt{3}-1}{4} = \frac{1}{2}.$$

Using Pythagoras' theorem, we find

$$D_{12} = \sqrt{\Delta x^2 + \Delta y^2} = 1.$$

Similar approach can be used to find the distance between the points (1) and (4), with the result $D_{14} = 1$. The figure is a square with a side 1 and the area 1.

The conclusion of this exercise is that the choice of the coordinate system affects the difficulty of the problem. For the same figure (a square) the case (C) is evidently more complicated than (A). Sometimes it is worth considering the same problem in different coordinates systems.

Problem 2.3

The relation between the Cartesian and polar coordinates is given by

$$x = r\cos\phi,$$

$$y = r\sin\phi.$$

Squaring both equations and adding them, we get

$$r^2 = x^2 + y^2 \quad \rightarrow \quad r = +\sqrt{x^2 + y^2}. \tag{B.5}$$

The angle ϕ requires more careful approach. When $r = 0$, we immediately recognize the point as the pole and can write its coordinates—$(0, 0)$. From the formula for the distance r follows that it equals zero only when both x and y are zero. If the distance is not zero, then we may talk unambiguously about the angle ϕ. Three cases are possible:

1. If $x = 0$, $y \neq 0$, then then ϕ equals $\pi/2$ for $y > 0$ and $-\pi/2$ for $y < 0$. In short, $\phi = \text{sign}(y)\pi/2$.
2. If $x \neq 0$, $y = 0$, then ϕ equals 0 for $x > 0$ and π for $x < 0$. In short, $\phi = (1 - \text{sign}(x))\pi/2$.

3. If $x \neq 0$, $y \neq 0$, then it is possible to make a ratio $y/x = \tan \phi$, from which we find $\phi = \arctan(y/x)$.

Problem 2.4

Starting with the transformation equations

$$x' = x \cos \theta + y \sin \theta, \tag{B.6}$$

$$y' = -x \sin \theta + y \cos \theta, \tag{B.7}$$

we first multiply (B.6) by $\sin \theta$, and (B.7) by $\cos \theta$ to get

$$x' \sin \theta = x \cos \theta \sin \theta + y \sin^2 \theta, \tag{B.8}$$
$$y' \cos \theta = -x \sin \theta \cos \theta + y \cos^2 \theta. \tag{B.9}$$

Adding up the last two equations results in

$$y = x' \sin \theta + y' \cos \theta.$$

Next, multiplying (B.6) by $\cos \theta$, and (B.7) by $\sin \theta$ we obtain

$$x' \cos \theta = x \cos^2 \theta + y \sin \theta \cos \theta, \tag{B.10}$$
$$y' \sin \theta = -x \sin^2 \theta + y \cos \theta \sin \theta. \tag{B.11}$$

Subtracting up the latter from the former results in the second transformation equation

$$x = x' \cos \theta + y' \sin \theta.$$

The answer could also be obtained quicker, once we notice that the rotation of the coordinate system $S' = (O', x', y')$ by the angle θ relative to $S = (O, x, y)$ is equivalent to the rotation of the S by the angle $-\theta$ relative to S'. Substituting θ with $-\theta$ in the original Eqs. (B.6) and (B.7) we can immediately find the desired expressions.

Problem 2.5

Using the transformation equations

$$x' = x \cos \theta + y \sin \theta, \tag{B.12}$$

$$y' = -x \sin \theta + y \cos \theta, \tag{B.13}$$

we first obtain

$$x'_Q - x'_P = (x_Q - x_P)\cos\theta + (y_Q - y_P)\sin\theta,$$

and

$$y'_Q - y'_P = -(x_Q - x_P)\sin\theta + (y_Q - y_P)\cos\theta.$$

Denoting

$$\Delta x = (x_Q - x_P),$$
$$\Delta x' = (x'_Q - x'_P),$$
$$\Delta y = (y_Q - y_P),$$
$$\Delta y' = (y'_Q - y'_P),$$

we can write

$$\Delta x'^2 = \Delta x^2 \cos\theta^2 + \Delta y^2 \sin\theta^2 + 2\Delta x \Delta y \cos\theta \sin\theta,$$

and

$$\Delta y'^2 = \Delta x^2 \sin\theta^2 + \Delta y^2 \cos\theta^2 - 2\Delta x \Delta y \cos\theta \sin\theta.$$

Summation of the last two equations results in

$$\Delta x'^2 + \Delta y'^2 = \Delta x^2 + \Delta y^2,$$

which proves that the distance equation in Cartesian coordinates is the same for S' and S.

Problem 2.6

Quadratic equation in variable v

$$xv^2 + yv + z = 0$$

has only one solution if the discriminant $D = y^2 - 4xz$ equals zero:

$$y^2 - 4xz = 0 \quad \rightarrow \quad z = \frac{y^2}{4x}.$$

The set of points (figure) defined by the equation

$$z = \frac{y^2}{4x}$$

is possible to construct by studying several "slices". First, if we consider the slices for various fixed distances along the x axis, we will get a set of parabolas

$$z = A_x y^2,$$

where A_x is getting smaller for larger values of x.

Second, considering slices for various fixed distances along the y axis, we get a set of multiplicative inverse functions

$$z = \frac{B_y}{x},$$

where B_y is getting larger for larger values of y.

Combining slices into a $3D$ picture, we arrive at the surface shown in Fig. B.16.

Problem 2.7

Going from the state 1 to 2, the gas occupies the same volume, while the pressure decreases. From the equation of state

$$PV = NkT$$

follows that the drop in the pressure is related to the decrease in temperature. The gas is getting colder.

Going from the state 2 to 3, the pressure of the gas remains the same, while the volume is increasing. From the equation of state follows that the expansion of the gas is related to the increase of temperature. The gas is getting hotter and expands.

Going from the state 3 to 1, the pressure of the gas increases, and the volume decreases. The temperature can not remain constant during this process, because that would require the curve P vs. V to look like multiplicative inverse function

$$P = \frac{NkT}{V},$$

while the curve connecting the states is a straight line.

The temperatures are easily found from the equation of state:

$$T_1 = \frac{P_1 V_1}{Nk},$$

$$T_2 = \frac{P_2 V_2}{Nk},$$

$$T_3 = \frac{P_3 V_3}{Nk}.$$

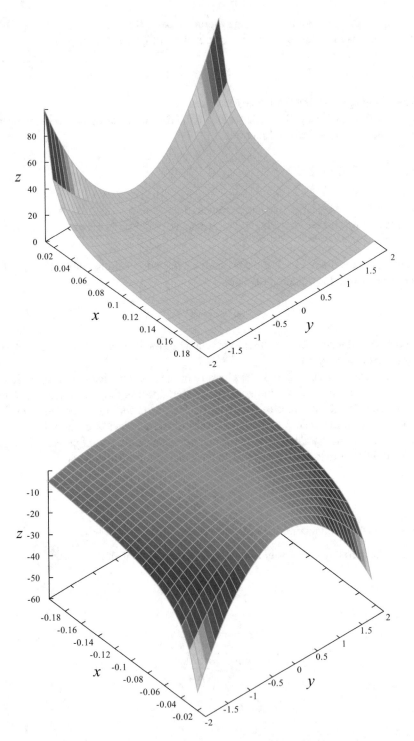

Fig. B.16 Figure in polynomial space defined by the requirement that the polynomial equation $P(v) = xv^2 + yv + z = 0$ has only one solution

B.18   Galilean Relativity

Problem 4.1

The problem is formulated in the reference frame of the ground, which we denote S; see Fig. B.17a. Let us switch to the reference frame of the wall—S'. The wall is moving with constant speed and can be used as an inertial reference frame.

In the frame of reference of the wall the ball is moving with the speed

$$w = v + u$$

towards to wall. The collision is elastic and the ball bounces off the wall with the same speed (relative to the wall), simply reversing the direction of its motion, see Fig. B.17b,c.

Since we are interested in the ball's final speed relative to the ground, we need to switch back to the reference frame S. The wall is moving with the sped u relative to the ground and the ball is moving with the speed $v + u$ relative to the wall, therefore the ball is moving with the speed

$$v' = v + 2u$$

relative to the ground.

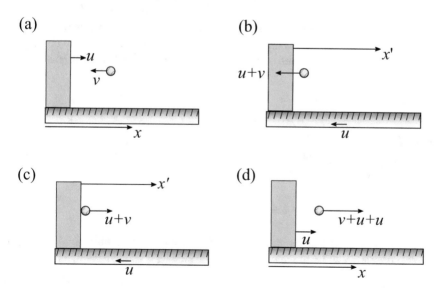

Fig. B.17 An elastic ball collides with a massive wall. (**a**) The problem is formulated in the reference frame of the ground. (**b**) After switching into the reference frame of the wall, the ball receives a "boost"; the ground is moving with the speed u, relative to the wall. (**c**) The ball collides with the wall elastically and flips the direction of its motion, keeping the speed the same. (**d**) Switching back into the reference frame of the ground; the ball received a "boost" again

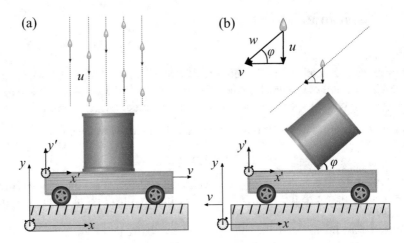

Fig. B.18 A bucket on a moving platform under the rain. (**a**) In the reference frame of the ground the rain-drops are falling vertically with the speed u, the cart is moving horizontally with the speed v. (**b**) In the reference frame of the cart, each rain-drop receives the horizontal component for its velocity, making the trajectory slanted

Problem 4.2

The problem is stated in the reference frame of the ground. Let us consider it in the inertial reference frame of the platform. In this reference frame the ground and the rain will have an additional horizontal velocity v, as shown in Fig. B.18b. Total velocity of the rain droplets is the result of vector addition of the velocity of the rain (relative to the ground) and the negative velocity of the platform (also relative to the ground).

$$\mathbf{w} = -\mathbf{v} + \mathbf{u}.$$

From Fig. B.18b it can been seen that the velocities $\mathbf{w}$, $\mathbf{u}$ and $-\mathbf{v}$ form a right triangle, with the hypotenuse w making an angle ϕ with the horizontal axis. From the definition of $\tan \phi$ follows:

$$\tan \phi = \frac{u}{v} \quad \rightarrow \quad \phi = \tan^{-1} \frac{u}{v}.$$

If the bucket is rotated so that its side is making an angle ϕ with horizontal axis, then the rain droplets will be moving parallel to the sidewalls of the bucket, as required.

Problem 4.3

In the reference frame of the unstable particles, the product particles can move in any direction, as long as their velocities point in opposite directions (to conserve their total momentum). This is illustrated in Fig. B.19a. The heads of the arrows $\mathbf{u}$ describe a circle of the radius u. To find the trajectory of a particle in the laboratory

Fig. B.19 (a) Product particles have total velocities due to their motion in the reference frame of the original unstable particle, plus the motion of the latter: $\mathbf{w} = \mathbf{u} + \mathbf{v}$. (b) For a certain pair of product particles, the resulting trajectory of one of them will have the maximal angle α, corresponding to the velocity $\mathbf{w}_1$ tangent to the circle of all possible directions of the velocities $\mathbf{u}$

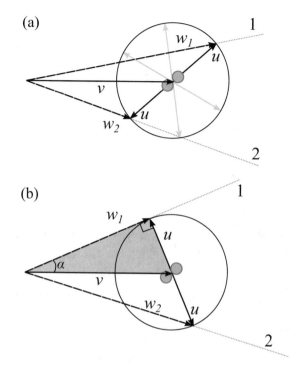

reference frame, we must add the velocity of the unstable particle $\mathbf{v}$ and the velocity of a product particle $\mathbf{u}$ relative to the unstable particle. The results, $\mathbf{w}_1$ and $\mathbf{w}_2$, for an arbitrary (not the one we are looking for!) pair of product particles is shown in Fig. B.19a. The lines 1 and 2 demonstrate the trajectories of the product particles in the laboratory reference frame.

For a certain pair of product particles, the trajectory of one of them will have the maximum angle relative to the direction of the original particle, as Fig. B.19b shows. For this trajectory, the velocity $\mathbf{w}_1$ is tangent to the circle of all possible orientations of $\mathbf{u}$. From the shaded triangle, with $\mathbf{w}_1 \perp \mathbf{u}$, we can find

$$\sin \alpha = \frac{u}{v}.$$

Problem 4.4

The problem is formulated in the reference frame of the ground. However, it is helpful to analyze it in the reference frame of one of the cars. Figure B.20 corresponds to the reference frame of the left car.

In the reference frame of the left car the ground is moving with the velocity $-\mathbf{v}$. The total velocity of the other car is then given by the sum of its velocity relative to the ground $\mathbf{u}$ and the velocity of the ground relative to the first car. The components if the velocity $\mathbf{w}$ are easily found:

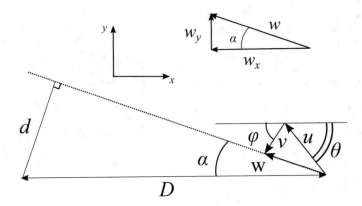

Fig. B.20 In the reference frame of the first car (the left one), the trajectory of the second car is determined by the total velocity $\mathbf{w} = \mathbf{u} - \mathbf{v}$. To find the angle α we can use the "velocity" triangle and the lengths of each side: $\tan \alpha = w_y / w_x$

$$w_x = -u \cos \theta - v \cos \phi = -(u \cos \theta + v \cos \phi),$$

and

$$w_y = u \sin \theta - v \sin \phi.$$

The direction of the trajectory of the right car is determined by the angle α, such that

$$\tan \alpha = \frac{w_y}{w_x} = \frac{v \sin \phi - u \sin \theta}{u \cos \theta + v \cos \phi}.$$

The minimal distance between the cars is determined by the shortest line segment from the location of the left car and the trajectory of the right car—the length of the perpendicular:

$$d = D \cos \alpha.$$

Using the trigonometric identity

$$\cos^2 \alpha = \frac{1}{1 + \tan^2 \alpha}$$

we can write

$$d = \frac{D}{\sqrt{1 + w_y^2/w_x^2}} = D \frac{|w_x|}{\sqrt{w_x^2 + w_y^2}}.$$

Using the expressions for the components w_x and w_y, after some simplifications, we obtain

$$d = D \frac{v \cos \phi + u \cos \theta}{\sqrt{v^2 + u^2 - 2uv \cos(\theta - \phi)}}.$$

Problem 4.5

The problem is best solved in the reference frame of the ground. The time required for the walls to close is

$$T = \frac{D}{2v}.$$

During this time the cockroach is running with the same speed (but not the same velocity!) Therefore, the total distance traveled is

$$L = uT = 2D \frac{v}{u}.$$

B.19 Spacetime

Problem 3.1

Many different approaches to synchronization of remote identical clocks are possible. Figure B.21 illustrates several methods.

Fig. B.21 Three methods of synchronization of remote clocks: (**a**) Transport of time from A to B; (**b**) using the symmetric signaling; (**c**) Using round-trip time (Poincare-Einstein synchronization)

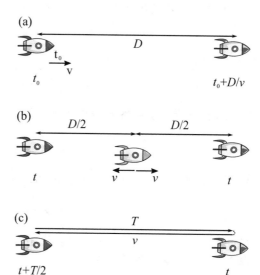

The first approach is based on sending a "times-stamped" message from one observer to the other. The observer A sends a message at the moment of time t_0 on its clock and this time is encoded in the message. When the message is received by the observer B, its clock is set to

$$t = t_0 + \frac{D}{v},$$

where D is the distance between the observers, v is the speed of the time-carrying signal. Both D and v should be independently known in this approach.

The second approach, shown in Fig. B.21b, requires a source of a signal placed in the middle between the observes A and B. This is done by placing the source C at such a position that the round-trip of a light-signal from A to B is double of the round-trip from A to C; or, alternatively, the round-trip times from C to A and from C to B are equal. This way, the knowledge of the distance D is not required. Once the source C is placed, it emits a flash of light propagating in all directions. When the light from C reaches the observers A or B, they set their clocks to a pre-arranged value, for example $t = 0$. Note that in this approach the value of the speed of light is not used.

Finally, the Poincare-Einstein synchronization approach can be used. It involves sending a signal from the observer A to B, where it is immediately reflected and received back at A after time T—the round-trip time. A and B agree to set their clocks as follows: When B receives a signal it sets its clock to a pre-arranged value t (e.g. zero); when A receives the reflected signal, it sets its clock to $t + T/2$.

Problem 3.2

The relationship between the time of emission $t_E = t_1$, the time of detection $t_D = t_2$, and the time t and position x of the reflection event is expressed as follows:

$$t = \frac{t_2 + t_1}{2},$$

$$x = \frac{t_2 - t_1}{2}.$$

The sum of these equations results in

$$t_2 = t + x,$$

while the subtraction of the second equation from the first yields

$$t_1 = t - x.$$

Note that in the above, x should be understood as the *distance* to the event. The more appropriate formulas are

$$t_1 = t - |x|, \quad t_2 = t + |x|.$$

Fig. B.22 Spacetime
diagram of two signals
emitted at different times but
reflected at the same time
from objects at different
distances away from the radar

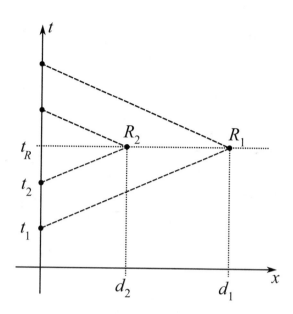

This is due to the fact that from the times of the radar t_1 and t_2, such that $t_2 > t_1$ one can not determine the direction in which the located event happened, only the distance to it.

Problem 3.3
The spacetime diagram is shown in Fig. B.22. Two reflections from objects at different locations can happen simultaneously. The reflection events R_1 and R_2 will be assigned the same time t_R, even though the reflected signal will not be received at the same time.

Problem 3.4
Figure B.23 demonstrates the solutions for all three cases.

When the object is at rest relative to the radar, the times between the emission, reflection, and detection events are all equal:

$$t_{E_2} - t_{E_1} = t_{R_2} - t_{R_1} = t_{D_2} - t_{D_1}.$$

As the object is approaching the radar, each subsequent signal has to cover less distance than the previous, thus the times between the emission, reflection, and detection events are related as follows:

$$t_{E_2} - t_{E_1} > t_{R_2} - t_{R_1} > t_{D_2} - t_{D_1}.$$

As the object is receding from the radar, each subsequent signal has to cover greater distance than the previous, thus the times between the emission, reflection,

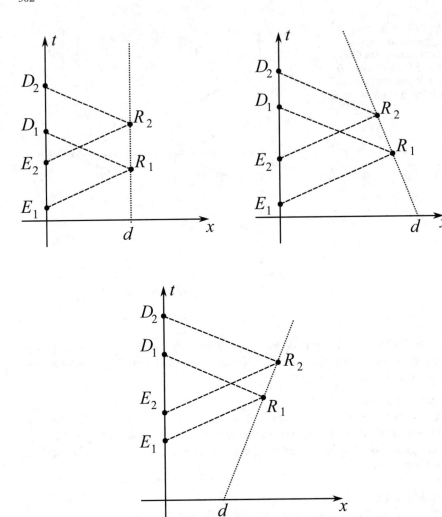

Fig. B.23 Spacetime diagrams for radar method used to locate an object in three cases: (**a**) Object at rest; (**b**) object approaching; (**c**) object receding

and detection events are related as follows:

$$t_{E_2} - t_{E_1} < t_{R_2} - t_{R_1} < t_{D_2} - t_{D_1}.$$

Problem 3.5

The spacetime diagram is shown in Fig. B.24. As the diagram illustrates, the observer A will first "see" the clock of the observer B ticking faster than its own. Then, after the observers meet, the clock of B will visually appear to be ticking

Fig. B.24 The observer B
sends images of its clock
every T seconds towards the
observer A. The latter at first
receives ("sees") the images
separated by a shorter time
T/k; after the meeting the
images are received separated
by a longer time kT

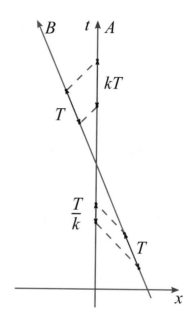

slower, compared to the clock of A. These are, respectively, the relativistic blue-shift and red-shift effects.

B.20   Special Relativity

Problem 5.1

Acceleration is defined as the change of velocity $\Delta \bar{v}$ during a time interval $\Delta \bar{t}$:

$$\bar{a} = \frac{\Delta \bar{v}}{\Delta \bar{t}} = \frac{\bar{c} \Delta v}{\Delta t} = \bar{c} a.$$

Therefore, the acceleration

$$a = \frac{\Delta v}{\Delta t} = \frac{\bar{a}}{\bar{c}}$$

has the units of inverse seconds, or Hertz.

Problem 5.2

The solution is shown in Fig. B.25. It is important to remember: the speed of the observer A relative to B is the same as the speed of B relative to A, thus the slope of the A's worldline is similar to the slope of the B's worldline in Minkowski diagram for the A's reference frame. Also, since every observer measures the speed

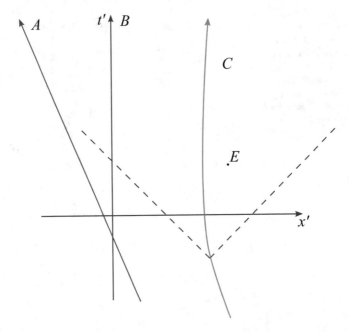

Fig. B.25 Minkowski diagram for events from Fig. 5.2, drawn for the reference frame *B*

of light to be 1 second per second, the worldlines of light are diagonal lines, going symmetrically between the spatial and temporal axes for all observers.

Problem 5.3

One way classical physics can answer this questions is to assume that there is no proper void, and whole space is filled with some medium, similar to ether. Modern field theories, such as Quantum Field Theory, express a similar idea. For example, the absence of electromagnetic radiation does not mean the absence of electromagnetic field. The field, which extends over the whole space, is likened to a stretched string: Waves are periodic excitations of the string. When there are no excitations, there is still "a string"—a minimal (ground) state of the field. Additionally, there is no true rest for a "string", there are always fluctuations above the ground state.

Step-by-Step Problem: Relativity of Simultaneity

The diagram for Part A is shown in Fig. B.26. The diagram for Part B is shown in Fig. B.27

Problem 5.4

Minkowski diagram is shown in Fig. B.28. Let us denote as t_1' and t_2' the times of emission of the first and the second light pulses, respectively, *according to B's clock.* Also, denote as T_1' and T_2' the times of the reception of corresponding signals.

Fig. B.26 Clock
synchronization for two
observers, A and B, using a
light signal emitted from the
middle. The reference frame
of the observer in the middle
is used

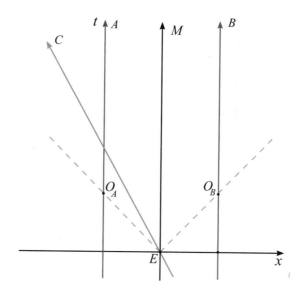

Fig. B.27 Clock
synchronization for two
observers, A and B, using a
light signal emitted from the
middle. The reference frame
of the observer C is used

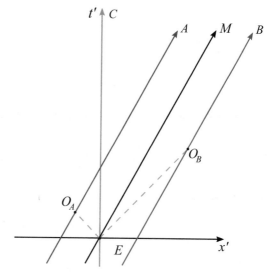

If at the moment of emission of the first pulse, the distance between the observer
B and A is D_1, then the signal will arrive to B at the moment

$$T_1' = t_1' + D_1/c = t_1' + D_1.$$

Similarly,

$$T_2' = t_2' + D_2.$$

Fig. B.28 Minkowski
diagram for two receding
observers, A and B, drawn for
the reference frame of B. The
observer A sends a pair of
signals towards B. According
to B, the interval between the
received signals is greater
than the interval between the
emitted signals: $\Delta T' > \Delta t'$

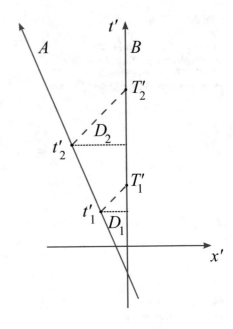

Therefore,

$$T_R^{(B)} = T_2' - T_1' = t_2' - t_1' + (D_2 - D_1) = T_E^{(B)} + \Delta D.$$

Between the emission of the first and the second pulses the observer A will move
by

$$\Delta D = v(t_2' - t_1') = v T_E^{(B)},$$

which leads to

$$T_R^{(B)} = (1 + v) T_E^{(B)}.$$

Denoting

$$k^{(B)} = 1 + v,$$

we can write

$$T_R^{(B)} = k^{(B)} T_E^{(B)}, \quad k^{(B)} \geq 1.$$

Problem 5.5
Figure B.29 shows Minkowski diagram for the reference frame of the observer B.
If the first signal from B is emitted at time t_B of its clock, then it will arrive back at

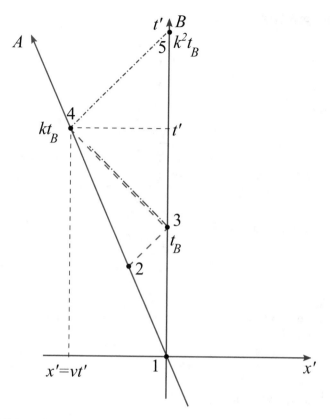

Fig. B.29 Minkowski diagram of signal exchanges between the observers A and B, used to locate each other and determine the expression of the relativistic Doppler k-factor. The reference frame of the observer B is used

$k^2 t_B$. The spacetime coordinates of the event 4 is then

$$t' = \frac{k^2 t_B + t_B}{2} = t_B \frac{k^2 + 1}{2},$$

and

$$x' = \frac{k^2 t_B - t_B}{2} = t_B \frac{k^2 - 1}{2}.$$

The speed of the observer A relative to B is

$$v = \frac{x'}{t'} = \frac{k^2 - 1}{k^2 + 1}.$$

Solving for k results in

$$k = \sqrt{\frac{1+v}{1-v}}.$$

Problem 5.6
Starting with

$$k^2 = \frac{1+v}{1-v},$$

and multiplying both sides by $1 - v$, we get

$$k^2 - vk^2 = 1 + v, \quad \rightarrow \quad k^2 - 1 = vk^2 + v.$$

From the last expression follows

$$v = \frac{k^2 - 1}{k^2 + 1}.$$

Problem 5.7
When the relative velocity of two observers is small, the value of the k factor is approximated well by

$$k = 1 + v,$$

from which immediately follows

$$z = k - 1 = v.$$

Problem 5.8
Firstly, when the scenario is "played backwards in time", the detector becomes the emitter (the observer B becomes the emitter). Secondly, the emitter becomes the detector (the observer A becomes the detector). Finally, the time interval between the received signals is shorter by the same factor k. In other words, when two observers are approaching each other, the time intervals between emitted and received signals are related as

$$T_R = \frac{T_E}{k},$$

where k is Doppler factor for the original scenario of receding observers.

Problem 5.9
From the definition of the relative shift

$$z = \frac{\lambda_R - \lambda_E}{\lambda_E} = \frac{\lambda_R}{\lambda_E} - 1 = k - 1,$$

follows

$$k = z + 1.$$

Using the relationship between the relative speed v and k

$$v = \frac{k^2 - 1}{k^2 + 1},$$

we can write the relative speed in terms of the shift z:

$$v = \frac{(z + 1)^2 - 1}{(z + 1)^2 + 1}.$$

For $z = 0.01$ the relative speed is $v = 0.01$; for $z = 0.1$, $v = 0.095$; for $z = 1.0$, $v = 0.6$; for $z = 10.0$, $v = 0.98$.

Problem 5.10
Using the expression for the Doppler k-factor in terms of the received and emitted wavelengths

$$k = \frac{T_R}{T_E} = \frac{\lambda_R}{\lambda_E},$$

we find

$$k = 1.076,$$

and corresponding velocity

$$v = \frac{k^2 - 1}{k^2 + 1} = 0.073.$$

In meter-based units it equals

$$\bar{v} = v\bar{c} = 22 \times 10^3 \text{ km/s}.$$

Problem 5.11
The time interval between the events of departure from the Earth and arrival at Alpha Centauri, as measured by the spaceship, is related to the time measured by the observer on Earth

$$T_{earth} = \gamma T_{ship} = \frac{T_{ship}}{\sqrt{1 - v^2}}.$$

In the reference frame of the Earth, the travel time for the ship is

$$T_{earth} = \frac{d}{v}.$$

Combining the last two equations, we obtain

$$\frac{d}{v} = \frac{T_{ship}}{\sqrt{1 - v^2}}.$$

Solving for v results in

$$v = \frac{1}{\sqrt{1 + (T_{ship}/d)^2}}.$$

Plugging in numerical values, we find $v = 0.9994$.

Problem 5.12
Figure B.30 shows Minkowski diagram of the events involved in the muon experiment. Two laboratories are moving with the speed close to the speed of light relative to the stationary muon. When the first laboratory (the one at the top of the mountain) is passing by the muon, both set their clocks to zero. When the second laboratory (the one at the bottom) is passing by the muon, the clock of the latter measures time T'. We can find the time of this event, as measured by the first

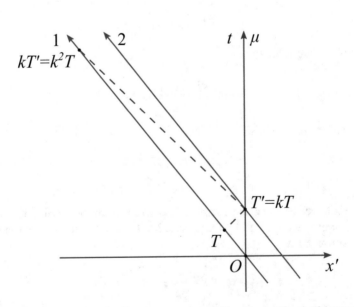

Fig. B.30 Minkowski diagram for the muon experiment. The reference frame of a muon is used

laboratory, as follows. Imagine a radar method is used by the first laboratory to locate the event when the muon and the second laboratory meet. If the locating signal is emitted at time T, according to the clock of the first laboratory, then the signal will be reflected at $T' = kT$, according to the clock in the muon's reference frame. The reflected signal will be received at time $kT' = k^2 T$ in the reference frame of the first laboratory. Therefore, the reflection happens at time

$$t_R = \frac{T + k^2 T}{2} = T\frac{k^2 + 1}{2}.$$

Since the time of this event in the muon's reference frame is $T' = kT$, we have

$$\frac{t_R}{T'} = \frac{1}{2}\left(k + \frac{1}{k}\right).$$

Using the expression for the Doppler factor k in terms of velocity, we can show that (see Problem 5.14):

$$\frac{t_R}{T'} = \gamma > 1.$$

Problem 5.13
Although it is possible to solve this problem by repeatedly applying the velocity composition formula

$$w = \frac{v + u}{1 + uv},$$

the solution turns out quite difficult. A simpler solution is found once we remember that instead of composing velocities, we can find the k factor that quantifies the relative motion of two reference frames. It was demonstrated that if C is moving relative to B with the k_1 factor, while B is moving relative to A with the k_2 factor, then C is moving relative to A with the $k_3 = k_1 k_2$ factor. Taking this into account, we immediately find that the N-th cart is moving relative to the ground with the k_N factor given by

$$k_N = k^N,$$

where

$$k = \sqrt{\frac{1 + v}{1 - v}}.$$

The relationship between k_N and corresponding velocity is given by (see Problem 5.6)

$$w = \frac{k_N^2 - 1}{k_N^2 + 1},$$

from which follows

$$w = \frac{(1 + v)^N - (1 - v)^N}{(1 + v)^N + (1 - v)^N}.$$

Problem 5.14

The expression for the relativistic Doppler factor k is given by

$$k = \sqrt{\frac{1 + v}{1 - v}}.$$

It can be re-written in two different ways:

$$k = \gamma(1 + v),$$

$$k = \frac{1}{\gamma(1 - v)}.$$

From these follow two equalities:

$$1 + v = \frac{k}{\gamma},$$

$$1 - v = \frac{1}{k\gamma}.$$

The sum of the last two equations results in

$$2 = \frac{k}{\gamma} + \frac{1}{k\gamma},$$

from which one can find γ:

$$\gamma = \frac{1}{2}\left(k + \frac{1}{k}\right).$$

Problem 5.15

From the spacetime diagram in Fig. 5.17 it is clear that the observer A will assign a time to the meeting of B and C half-way between the departure of B and the arrival of C:

$$t = \frac{T_A}{2} = \frac{T}{2}\left(k + \frac{1}{k}\right).$$

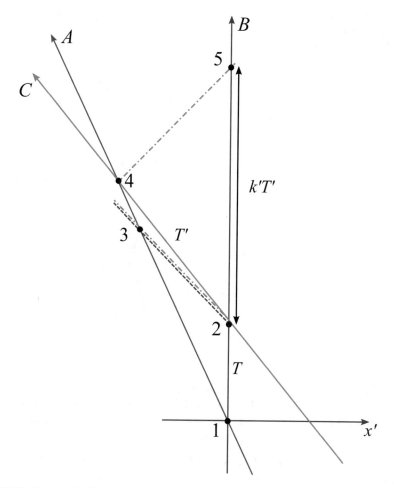

Fig. B.31 Minkowski diagram of the round-trip experiment demonstrating the path-dependent nature of time measurement. The reference frame of the inertial observer B is used

Using the result of the Problem 5.14, we can write

$$t = \gamma T > T.$$

Problem 5.16

(a): The Minkowski diagram, shown in Fig. B.31, is plotted for the reference frame of the observer B. Given that the observers A and B are receding, while the observers A and C are approaching, the time between the events 1 and 4, as measured by A equals

Fig. B.32 The relationship
between the Doppler k factors
for the observers A, B, and C

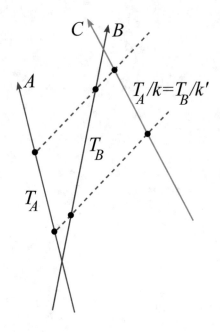

$$t_A = kT + \frac{T'}{k}.$$

It is important to remember that the relative motion of A and B, as well as of
A and C, is characterized by the same k factor. The relative motion of B and C
is characterized by a different Doppler factor k'. To find it, we can take a look
at the pair of signals emitted by A and received by B and C; see Fig. B.32.

Since the observers A and C are approaching, the time between the emitted
signals T_A is measured by C as

$$T_C = \frac{T_A}{k}.$$

Similarly for the approaching observers C and B:

$$T_C = \frac{T_B}{k'}.$$

The observers A and B are receding:

$$T_B = kT_A.$$

Therefore,

$$T_C = \frac{T_A}{k} = \frac{T_B}{k'} = \frac{kT_A}{k'} = \frac{T_A}{k},$$

from which follows

$$k' = k^2.$$

Using this result, we can complete the analysis of the Minkowski diagram in Fig. B.31.

The time on the C's clock at the event 4 is

$$t_C = T + T',$$

while the observer A will measure

$$t_A = kT + \frac{T'}{k}.$$

To find the relationship between T and T', we can write the time of an additional event 5. On the one hand,

$$t_5 = T + k'T' = T + k^2T';$$

on the other hand, it equals

$$t_5 = kt_A = k^2T + T'.$$

From the last two equations follows

$$T = T'.$$

This leads to

$$t_C = 2T,$$

and

$$t_A = kT + \frac{T}{k} = T\left(k + \frac{1}{k}\right).$$

We obtained the same result as in the reference frame of A.

(b): Figure B.33 shows Minkowski diagram for the reference frame of the observer C. Similar to the analysis performed above for the observer B, an additional light signal from the event 4–5 is introduced. The analysis for the reference frame of C is largely the same as for the reference frame of the observer B, and involves the comparison of various time intervals.

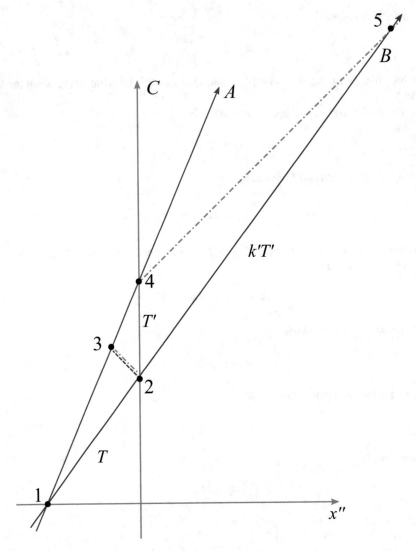

Fig. B.33 Minkowski diagram of the round-trip experiment demonstrating the path-dependent nature of time measurement. The reference frame of the inertial observer C is used

First, the time interval between the events 1 and 5 can be written as

$$t_5 = T + k'T' = T + k^2T',$$

or

$$t_5 = k(kT + \frac{T'}{k}) = k^2T + T'.$$

Again, we deduce that $T' = T$. The time at the event 4 on the C's clock is

$$t_C = T + T' = 2T,$$

while the clock of A will measure

$$t_A = kT + \frac{T'}{k} = T(k + \frac{1}{k}).$$

The analyses of the round-trip yield the same results, regardless of the reference frame.

Problem 5.17

The solution presented below uses an approach different from the one used to obtain the original formulas for the Lorentz transformation. The goal is to demonstrate an alternative way. We start with the Lorentz transformation

$$t' = \gamma(t - vx) = \gamma t - \gamma vx, \tag{B.14}$$

and

$$x' = \gamma(x - vt) = \gamma x - \gamma vt. \tag{B.15}$$

Multiplying (B.14) by v and adding to (B.15), we obtain

$$x' + vt' = \gamma x - \gamma v^2 x = \gamma(1 - v^2)x = \frac{x}{\gamma},$$

from which follows

$$x = \gamma(x' + vt').$$

Next, multiplying (B.15) by v and adding to (B.14), we obtain

$$t' + vx' = \gamma t - \gamma v^2 t = \gamma(1 - v^2)t = \frac{t}{\gamma}.$$

We then find

$$t = \gamma(t' + vx').$$

The form of the "direct" and "inverse" equations is the same; the difference in the sign in front of velocity reflects the fact that if B is moving with the velocity v relative to A, then A is moving with the velocity $-v$ relative to B.

Problem 5.18

We start with the Lorentz transformation

$$t' = \gamma(t - vx) = \frac{t - vx}{\sqrt{1 - v^2}}. \tag{B.16}$$

The time is measured in the same way in c-based and meter-based units. The velocity v is written as

$$v = \frac{\bar{v}}{c},$$

while the position x is

$$x = \frac{\bar{x}}{c}.$$

Plugging these equations into the formula for t', we get

$$\bar{t}' = \frac{\bar{t} - \bar{v}\bar{x}/\bar{c}^2}{\sqrt{1 - \bar{v}^2/\bar{c}^2}}. \tag{B.17}$$

The equation for x' is obtained in a similar way. The result is

$$\bar{x}' = \frac{\bar{x} - \bar{v}\bar{t}}{\sqrt{1 - \bar{v}^2/\bar{c}^2}}. \tag{B.18}$$

Problem 6.1

We demonstrated that if cause and effect events are connected with a causation signal propagating with the speed u_c (relative to A), then the observer B has to move relative to A with the speed v

$$v > \frac{1}{u_c}$$

to flip the order of this pair of events. If $u_c \to \infty$, the required speed becomes infinitely small. Motion with arbitrarily small relative speed v will be sufficient to flip the order *of any pair of events*. In other words, all cause-effect relations would loose their invariant character.

Problem 6.2

From the Lorentz transformation we have

$$\Delta t' = \gamma(\Delta t - v\Delta x).$$

In the reference frame A the events happen at the same time ($\Delta t = 0$.) Therefore,

$$\Delta t' = -\gamma v\Delta x.$$

The relative speed v is small, leading to γ almost exactly equal to unity. The time between the events, as measured by the observer B, is then

$$\Delta t' = -2573 \text{ years}.$$

The negative sign indicates that the second event happens first.

Problem 6.3

Figure B.34 shows Minkowski diagram of two identical rulers moving with different speeds. A worldline of a ruler at rest is also shown. Not only the slopes of the worldlines of two rulers are different, but their measured lengths are different as well:

$$L_1 = 0.95L_0, \quad L_2 = 0.6L_0.$$

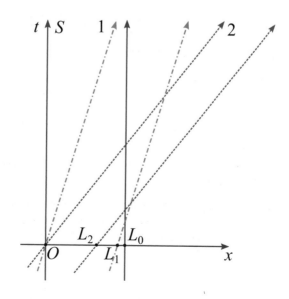

Fig. B.34 Two rulers with equal proper lengths L_0 are moving relative to a reference frame S with different velocities. The worldlines of their front and back have different slope; their measured lengths, L_1 and L_2, are also different

Problem 6.4

From the Lorentz transformation equations we first find

$$(\Delta t')^2 = \gamma^2[(\Delta t)^2 - 2v\Delta t \Delta x + v^2 \Delta x^2].$$

Next, we get

$$(\Delta x')^2 = \gamma^2[(\Delta x)^2 - 2v\Delta t \Delta x + v^2 \Delta x^2].$$

Subtracting the second from the first results in

$$(\Delta t')^2 - (\Delta x')^2 = \gamma^2[(\Delta t)^2 - v^2(\Delta t)^2 + v(\Delta x)^2 - (\Delta x)^2].$$

Simplifying the last expression, we arrive at

$$(\Delta t')^2 - (\Delta x')^2 = \gamma^2(1 - v^2)[(\Delta t)^2 - (\Delta x)^2] = (\Delta t)^2 - (\Delta x)^2.$$

Problem 6.5

Two key events in this problem are: E—the birth of a muon, and F—its decay. In the reference frame of the muon the interval squared is easily calculated:

$$\Delta s^2 = T_0^2.$$

If the muon is moving with the speed v relative to an observer S, it will decay after time T, traveling the distance $D = vT$. The expression for the interval squared in the reference frame S is

$$\Delta s^2 = T^2 - D^2 = \frac{D^2}{v^2} - D^2.$$

Equating the last two expressions, we find

$$v = \frac{D}{\sqrt{D^2 + T_0^2}}.$$

Problem 6.6

The diagram in Fig. B.35 demonstrates a pair of events, E and F, separated by a time-like interval.

(a) The straight line connecting the events corresponds to a possible observer present at both events. The clock of such an observer will measure the time between the events, which will be the longest time among any other worldline connecting E and F.

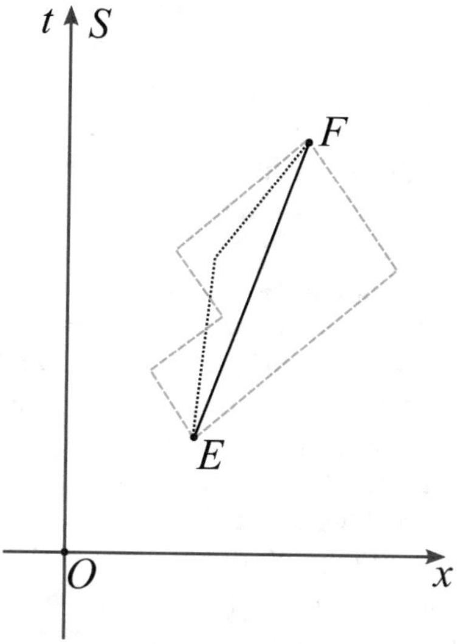

Fig. B.35 A pair of events, E and F, separated by a time-like interval, can be connected by many different "paths". The straight line represents the path of the longest spacetime length. Zig-zagged paths, plotted as dashed lines, correspond to light signal propagating (in a bouncing manner) between these events

Fig. B.36 Three types of "circles" in spacetime—points of constant spacetime distance from the origin. Solid blue lines correspond to the equation $t^2 - x^2 = 1$; dotted red lines correspond to the equation $t^2 - x^2 = -1$; dashed orange lines correspond to the equation $t^2 - x^2 = 0$. All these lines are *the same in all reference frames*

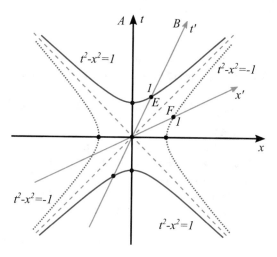

(b) The worldlines of light signal, connecting any pair of events, have zero space-time lengths. The diagrams shows two, among infinite number of possibilities, broken worldlines, describing propagation of light signal between the events E and F. The total length of each line is zero.

Problem 6.7

Figure B.36 shows Minkowski diagram for the reference frame of some observer A. Three types of curves are plotted, for constant positive interval squared (solid blue), for constant negative interval squared (dotted red), and for constant zero interval squared (dashed orange).

The equation $t^2 - x^2 = 1$ specifies all events that lie 1 s away from the origin, either in the future (top solid blue line), or in the past (bottom solid blue line) relative to the origin event. Every event on these curves is connected with the origin event with a time-like interval ($t^2 = x^2 + 1$, therefore $|t| > |x|$). An observer B, moving relative to A and present at the event E, will measure the proper time of 1 s between the meeting with A and the event E.

The equation $t^2 - x^2 = -1$ specifies all events that lie 1 s away in space for different observers. Every event on these curves is connected with the origin event with a space-like interval ($x^2 = t^2 + 1$, therefore $|x| > |t|$). An observer B, moving relative to A and present at the event E, will measure the proper distance/length of 1 s between the meeting with A and the event F.

Finally, the equation $t^2 - x^2 = 0$ specifies all events where a light signal is present. These events correspond to the light signal emitted from the origin, as well as reaching the origin event from the past.

It is important to understand that *these curves remain the same in all reference frames*. For example, in the reference frame of the observer B the equation for the events 1 s away in time is given by same equation as for the observer A:

$$t'^2 - x'^2 = 1,$$

and the curve in the Minkowski diagram (t', x') has the same shape and orientation. The same applies to the other curves with the interval squared equal -1 and 0. *Thus, similar to the diagonal light-lines, the curves of constant time-like separation and constant space-like separation are also invariant between all Minkowski diagrams.*

Problem 6.8

The space-like separation of a pair of events, E_1 and E_2, implies that in any reference frame the separation in space Δx_{12} is greater than the separation in time Δt_{12}. Suppose the measurements of Δt_{12} and Δx_{12} are performed in the reference frame of an observer A. An observer B, moving relative to A with the velocity v, will measure the time interval between the events as

$$\Delta t'_{12} = \gamma(\Delta t_{12} - v\Delta x_{12}).$$

If the velocity v satisfies the condition

$$v = \frac{\Delta t_{12}}{\Delta x_{12}} = v_0 < 1,$$

then the events will be simultaneous for B. The required velocity is allowed, since it is less than the speed of light. Note, that for the velocities

$$v_0 < v < 1,$$

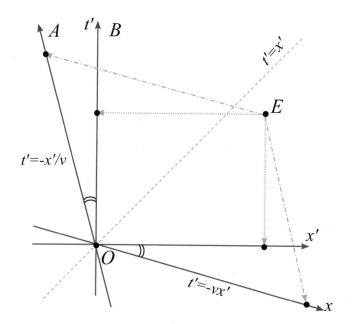

Fig. B.37 Time and space axes of the observer A in B's reference frame. All events simultaneous with $t = 0$ in the reference frame of A are given by the equation $t' = -vx'$

the events E_1 and E_2 will have the time order opposite to the order measured by the observer A.

Problem 6.9
Figure B.37 shows Minkowski diagram in the reference frame of the observer B. Relative to B, the observer A is moving with the velocity $-v$. The time coordinates of A and B are related:

$$t = \gamma(t' + vx').$$

The events with $t = 0$ all belong to the straight line

$$t' + vx' = 0, \quad \rightarrow \quad t' = -vx'.$$

This line is shown in Fig. B.37.

Problem 7.1
Using the results of the Problem 5.17, we immediately get

$$t = \gamma(t' + vx'),$$
$$x = \gamma(x' + vt'),$$

$$y = y',$$
$$z = z'.$$

Problem 7.2
Using the Lorentz transformation in the form

$$\Delta t' = \gamma(\Delta t - v\Delta x),$$
$$\Delta x' = \gamma(\Delta x - v\Delta t),$$
$$\Delta y' = \Delta y,$$
$$\Delta z' = \Delta z,$$

we can write

$$u'_x = \frac{\Delta x'}{\Delta t'} = \frac{\Delta x - v\Delta t}{\Delta t - v\Delta x},$$

which simplifies to

$$u'_x = \frac{u_x - v}{1 - vu_x},$$

where $u_x = \Delta x/\Delta t$. In a similar way we obtain,

$$u'_y = \frac{u_y}{\gamma(1 - vu_x)},$$

$$u'_z = \frac{u_z}{\gamma(1 - vu_x)}.$$

The only difference between the "direct" and the "inverse" transformations comes from the sign of relative velocity u_x.

Problem 7.3
The components of the photon's velocity relative to spacetime coordinates (t, x, y, z) are

$$u_x = \frac{v + \cos\alpha}{1 + v\cos\alpha},$$

$$u_y = \frac{\sin\alpha}{\gamma(1 + v\cos\alpha)},$$

and

$$u_z = u'_z = 0.$$

The magnitude squared is

$$u_x^2 + u_y^2 = \frac{\gamma^2(v + \cos\alpha)^2 + \sin^2\alpha}{\gamma^2(1 + v\cos\alpha)^2}.$$

The numerator of the last expression can be simplified, using the trigonometric identity $\sin^2\alpha = 1 - \cos^2\alpha$:

$$\gamma^2v^2 + \gamma^2\cos^2\alpha + 2\gamma^2v\cos\alpha + \sin^2\alpha$$
$$= \gamma^2v^2 + (\gamma^2 - 1)\cos^2\alpha + 2\gamma^2v\cos\alpha + 1.$$

Next, we will find that

$$\gamma^2 - 1 = \frac{1}{1 - v^2} - 1 = \frac{v^2}{1 - v^2} = \gamma^2v^2,$$

and write

$$\gamma^2v^2 + (\gamma^2 - 1)\cos\alpha^2 + 2\gamma^2v\cos\alpha + 1$$
$$= \gamma^2 - 1 + \gamma^2v^2\cos^2\alpha + 2\gamma^2v\cos\alpha + 1.$$

Finally, we arrive at

$$\gamma^2 + \gamma^2v^2\cos^2\alpha + 2\gamma^2v\cos\alpha$$
$$= \gamma^2(1 + 2v\cos\alpha + v^2\cos^2\alpha)$$
$$= \gamma^2(1 + v\cos\alpha)^2.$$

The same expression as for the denominator in the expression for $u_x^2 + u_y^2$. This proves that

$$u_x^2 + u_y^2 = 1.$$

Problem 7.4
The aberration formulas

$$\cos\theta' = \frac{v + \cos\theta}{1 + v\cos\theta},$$

and

$$\sin\theta' = \frac{\sin\theta}{\gamma(1 + v\cos\theta)},$$

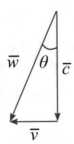

Fig. B.38 According to Galileo-Newtonian mechanics, in the reference frame of the Earth the particles of light propagate with the velocity $\bar{w}$, obtained by adding the velocity of light relative to the fixed stars $\bar{c}$ and the velocity $\bar{v}$, equal to the velocity of the Earth, but pointing in the opposite direction. In this approach, $\bar{w} > \bar{c}$—the result incompatible with the special theory of relativity

simplify for $\theta = 90°$ and $v \ll 1$ ($\gamma \approx 1$):

$$\cos \theta' = v, \quad \sin \theta' = 1.$$

Therefore,

$$\tan \theta' = \frac{1}{v} = \frac{\bar{c}}{v}.$$

The same answer is obtained from the analysis of velocity addition triangle, used in Galileo-Newtonian physics; see Fig. B.38.

Problem 7.5
Using the Lorentz transformation for the spacetime velocity

$$u'_t = \gamma (u_t - v u_x),$$

and

$$u'_x = \gamma (u_x - v u_t),$$

we obtain for $u_t = 1$ and $u_x = 0$:

$$u'_t = \gamma, \quad u'_x = -\gamma v.$$

Problem 7.6
The transformation equation

$$u'_x = \gamma (u_x - v u_t)$$

can be rewritten as follows

$$\gamma(u')u' = \gamma(v)[\gamma(u)u - v\gamma(u)]$$
$$= \gamma(v)\gamma(u)(u - v).$$

Denoting the right-hand side of the last equality as w, we obtain an equation for the unknown u':

$$\gamma(u')u' = w.$$

Solving for u' yields

$$(u')^2 = \frac{w^2}{1 + w^2} = \frac{1}{1 + 1/w^2}.$$

Substituting the expression

$$\frac{1}{w^2} = \frac{(1 - v^2)(1 - u^2)}{(u - v)^2},$$

after some simplification, we arrive at

$$(u')^2 = \frac{(u - v)^2}{(u - v)^2 + (1 - v^2)(1 - u^2)} = \frac{(u - v)^2}{(1 - uv)^2}.$$

Therefore,

$$u' = \frac{|u - v|}{(1 - uv)}.$$

Problem 7.7

Lorentz transformation for the components of the spacetime momentum are

$$p'_t = \gamma(p_t - vp_x),$$

$$p'_x = \gamma(p_x - vp_t).$$

Using the values $p_t = m$, $p_x = 0$, we obtain

$$p'_t = \gamma m, \quad p'_x = -\gamma mv.$$

Problem 7.8

Using the conservation laws, we write:

$$E + M = E'$$

for energy, and

$$E = P'$$

for momentum; here E' is the energy after the absorption, P'—momentum after the absorption. To find the mass of the atom we can use the relation between mass and energy:

$$E' = \gamma M'.$$

The value for the factor γ can be calculated if the speed of the atom after the absorption is known. The speed is easily found using the relationship between the momentum and energy:

$$P' = v E',$$

and the expressions for P' and E' given above.

$$v = \frac{P'}{E'} = \frac{E}{E + M}.$$

From the last expression we find

$$\gamma = \frac{E + M}{\sqrt{M(M + 2E)}}.$$

The mass of the atom after the absorption of the photon is given by

$$M' = \frac{E'}{\gamma} = \sqrt{M(M + 2E)}.$$

The last expression can be also written as

$$M' = M\sqrt{1 + 2E/M},$$

which, for small ratios E/M, is approximately equal to

$$M' \approx M + E.$$

To get the last expression we used the fact that

$$\sqrt{1 + x} \approx 1 + \frac{x}{2},$$

for $x \ll 1$.

Problem 7.9

First, we use the conservation laws:

$$M = E' + E,$$

for the energy, and

$$0 = P' - E$$

for momentum. From these we immediately find the energy $E' = M - E$ and momentum $P' = E$ of the atom after the emission of the photon.

From the relationship between energy and momentum

$$P' = vE',$$

we obtain the speed of the atom:

$$v = \frac{P'}{E'} = \frac{E}{M - E}.$$

Let us also find the mass of the atom after the emission of the photon. Using the speed of the atom we first find

$$\gamma = \frac{M - E}{\sqrt{M(M - 2E)}},$$

and then the mass M' from the relation $E' = \gamma M'$:

$$M' = \frac{E'}{\gamma} = \sqrt{M(M - 2E)}.$$

The last expression can be also written as

$$M' = M\sqrt{1 - 2E/M}.$$

When the energy of the emitted photon is much smaller than the rest energy of the atom (equal to its mass M) we can write

$$M' \approx M - E.$$

From the expressions for the speed and mass of the particle after the emission we can see that when $E = M/2$, the particle has to become massless and will be moving with speed $v = 1$. This is essentially a decay of a particle into two photons traveling in the opposite directions.

Problem 7.10

The conservation of energy requires that

$$E = 2m + M,$$

where $E = 2\gamma m$ is the energy of two protons before the collision, m denotes the proton's mass. The value for the factor γ immediately follows:

$$\gamma = \frac{M + 2m}{2m}. \tag{B.19}$$

From the definition of the factor γ in terms of velocity we obtain

$$v = \frac{\sqrt{\gamma^2 - 1}}{\gamma}. \tag{B.20}$$

Plugging in the expression for the factor γ (B.19), after some algebraic manipulations we arrive at

$$v = \frac{\sqrt{M(M + 4m)}}{M + 2m}.$$

If we rewrite formula (B.20) as

$$v = \sqrt{1 - \frac{1}{\gamma^2}},$$

then the expression for the proton's speed takes an alternative form:

$$v = \sqrt{1 - 1/(1 + M/2m)^2}.$$

Problem 7.11

From the conservation of energy we find

$$M = 2E,$$

where E is the energy of each photon. For massless particles $E = P$, therefore the momenta of each photon is

$$P = M/2.$$

Problem 7.12

When electron and positron annihilate their total energy (rest energy plus kinetic energy) is transformed into the energy of two protons. The energy required to produce two protons is minimal when both protons are at rest. Their total energy is purely rest energy. We thus can write the conservation of energy for this scenario:

$$2E = 2M,$$

where E is the energy of a moving electron (positron), M is the mass of a proton. The kinetic energy of electron (positron) can then be found:

$$E_k = E - m = M - m,$$

where m is the mass of electron (positron). The total kinetic energy of electron and positron is $2E_k$.

Problem 7.13

The conservation laws for this problem can be written as follows:

$$E = 2P,$$

for the energy;

$$Ev = P \cos \beta_1 + P \cos \beta_2,$$

for the horizontal momentum, and

$$0 = P \sin \beta_1 - P \sin \beta_2$$

for the vertical momentum. Here we used the relationship between the relativistic momentum and energy $P = Ev$.

From the conservation of momentum in the vertical direction follows that

$$\beta_1 = \beta_2 = \frac{\alpha}{2}.$$

Using the conservation of momentum in horizontal direction and the equality $Ev = 2Pv$ we arrive at

$$2Pv = 2P \cos \frac{\alpha}{2}.$$

The speed of the particle is then

$$v = \cos \frac{\alpha}{2}.$$

Problem 7.14

The speed of the particle before the collision can be found if its momentum and energy before the collision are known:

$$v = \frac{P_{before}}{E_{before}}.$$

The conservation of momentum requires that

$$P_{before} = P_{after} = E.$$

Writing the energy of a moving particle as

$$E_{before} = \gamma(v)M,$$

we arrive at

$$v = \frac{E}{\gamma(v)M}.$$

Rewriting the last expression as

$$v\gamma(v) = \frac{v}{\sqrt{1 - v^2}} = \frac{E}{M},$$

we can solve for v to find

$$v = \frac{E}{\sqrt{E^2 + M^2}}.$$

B.21 Additional Problems

Problem B.9
Let us first find the Cartesian coordinate x of any point, provided we know its oblique coordinates x' and y'. Figure B.39 illustrates the required steps.

The distance Ox_P is the sum of distances OQ and Qx_P. Since OQ is the side of the right triangle OQx'_P adjacent to the angle OQx'_P, we have

$$OQ = x'_P \cos\theta. \tag{B.21}$$

Similarly, since $Qx_P = x'_P R$ and $x'_P R$ is the side of the right triangle PRx'_P opposite to the angle RPx'_P, we may write

$$Qx_P = y'_P \sin\theta, \tag{B.22}$$

resulting in

$$x_P = x'_P \cos\theta + y'_P \sin\theta. \tag{B.23}$$

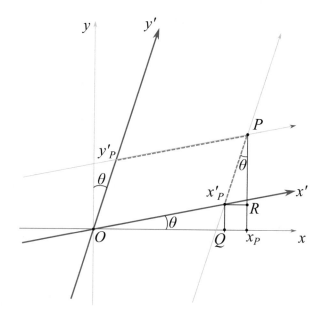

Fig. B.39 Expressing the Cartesian coordinate x of a point in terms of its oblique coordinates x' and y'. The distance from O to x_P is the sum of $OQ = x'_P \cos \theta$ and $Qx_P = y'_P \sin \theta$

Next, we can find the Cartesian coordinate y of a point in terms of its oblique coordinates x' and y'. From Fig. B.40 it is not difficult to see that the arguments very similar to the ones used for the x coordinates are applicable in this case.
The result is

$$y_P = y'_P \cos \theta + x'_P \sin \theta$$

or

$$y_P = x'_P \sin \theta + y'_P \cos \theta.$$

To summarize:

$$x = x' \cos \theta + y' \sin \theta,$$
$$y = x' \sin \theta + y' \cos \theta.$$

The inverse transformations are given by

$$x' = \frac{x \cos \theta - y \sin \theta}{\cos 2\theta},$$

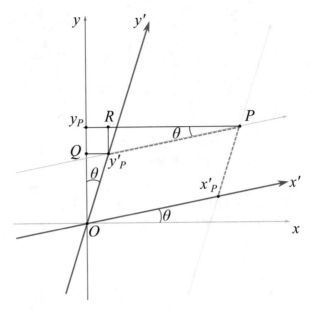

Fig. B.40 Expressing the Cartesian coordinate y of a point in terms of its oblique coordinates x' and y'. The distance Oy_P is given by the sum of $OQ = y'_P \cos \theta$ and $Qy_P = x'_P \sin \theta$

$$y' = \frac{-x \sin \theta + y \cos \theta}{\cos 2\theta}.$$

An interesting observation can be made about the expression for the distance squared in the oblique coordinates. Given two points, separated by $\Delta x'$ and $\Delta y'$ in oblique coordinates, we can find their separation along each axis in the Cartesian coordinates:

$$\Delta x = \Delta x' \cos \theta + \Delta y' \sin \theta,$$
$$\Delta y = \Delta x' \sin \theta + \Delta y' \cos \theta. \tag{B.24}$$

In Cartesian coordinates the distance squared is given by

$$d^2 = \Delta x^2 + \Delta y^2, \tag{B.25}$$

which, after the substitution of expressions (B.24), leads to

$$d^2 = \Delta x'^2 + 2\Delta x' \Delta y' \cos \phi + \Delta y'^2. \tag{B.26}$$

The angle

$$\phi = \pi/2 - 2\theta$$

Fig. B.41 Distance squared between two points can expressed in oblique coordinates using the law of cosines

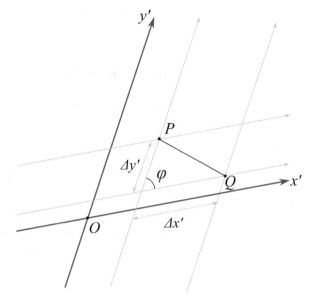

is the angle between the coordinate axes of the oblique coordinate system. This formulas represents the well known law of cosines, as illustrated in Fig. B.41.

The expression (B.26) for the distance squared is valid for both oblique coordinates and for Cartesian coordinates. The latter constitute a special case, when $\phi = \pi/2$, and the law of cosines reduces to Pythagoras' theorem.

Problem B.2

The relationships between the Cartesian coordinates of C' and C are

$$x' = x \cos \alpha + y \sin \alpha,$$
$$y' = -x \sin \alpha + y \cos \alpha.$$

Similar relationships apply for the Cartesian coordinates of C'' and C':

$$x'' = x' \cos \beta + y' \sin \beta,$$
$$y'' = -x' \sin \beta + y' \cos \beta.$$

Substituting the expressions for x' and y' in terms of x and y we get

$$x'' = (x \cos \alpha + y \sin \beta) \cos \beta + (-x \sin \alpha + y \cos \alpha) \sin \beta.$$

Opening the parentheses and rearranging the terms results in

$$x'' = x(\cos \alpha \cos \beta - \sin \alpha \sin \beta) + y(\sin \alpha \cos \beta + \sin \beta \cos \alpha).$$

On the other hand, the relationships between (x'', y'') and (x, y) must have the form

$$x'' = x \cos(\alpha + \beta) + y \sin(\alpha + \beta).$$

Comparing the last two expressions we see that

$$\cos(\alpha + \beta) = \cos \alpha \cos \beta - \sin \alpha \sin \beta,$$

and

$$\sin(\alpha + \beta) = \sin \alpha \cos \beta + \sin \beta \cos \alpha.$$

Problem B.3

The problem is formulated in the reference frame of the ground. There are three more inertial reference frames which can be used to analyze and solve the problem: the runner at the back; the runner at the front, and the coach. Let us explore all these possibilities.

A: Reference frame of the ground.

In the reference frame of the ground, the position of the front runner, the back runner, and the coach can be written as

$$x_f = L - vt,$$

$$x_b = vt,$$

$$x_c = L - ut.$$

The origin of the coordinate axis x is placed at the position of the back runner at the moment when the front runner and the coach meet. This moment of time is chosen as $t = 0$. The positive direction of the x axis is left to right, from the back runner to the coach.

Some time T later, the coach meets the back runner, meaning

$$x_b(T) = x_c(T) \quad \rightarrow \quad T = \frac{L}{v + u}.$$

The distance between the front and the back runner at this moment will give the desired final length:

$$L' = x_b(T) - x_f(T) = \frac{Lv}{v + u} - L + \frac{Lv}{v + u} = L\frac{v - u}{v + u}.$$

B: Reference frame of the back runner.

Next, consider the process in the reference frame of the back runner. Relative to her, the speed of the coach is

$$u' = v + u,$$

and the speed of the front runner after meeting the coach is $2v$. It will take time $T = L/u'$ for the coach to reach the back runner. During this time the front runner will travel the distance $D = 2vT$ from the initial position of the coach. Since the coach is L meters away when the front runner reaches her, the position of the front runner when the coach reaches the back runner is $D - L$ meters on the left. Therefore, the final length of the line of joggers is

$$L' = D - L = 2vT - L = \frac{2vL}{v + u} - L = L\frac{v - u}{u + v}.$$

C: Reference frame of the front runner.

In the reference frame of the front runner, right after she meets the coach, the speed of the back runner is $u' = 2v$. The speed of the coach is $v - u$, and the coach is moving away, to the right, relative to the front runner. The positions of the coach and the back runner are

$$x_b = -L + 2vt,$$

$$x_c = (v - u)t.$$

The coach meets the back runner when $x_b = x_c$:

$$-L + 2vT = (v - u)T \quad \rightarrow \quad T = \frac{L}{v + u}.$$

At this moment the back runner will be at

$$x_b(T) = -L + \frac{2vL}{v + u} = L\frac{v - u}{v + u},$$

relative to the front runner.
D: Reference frame of the coach.
Finally, in the reference frame of the coach the front runner is moving away with the speed $v - u$, while the back runner is approaching with the speed $v + u$. To cover the distance L between the coach and the back runner, the latter will require time

$$T = \frac{L}{v + u}.$$

By this time, the front runner will cover the distance

$$d = (v - u)T = L\frac{v - u}{v + u}.$$

As can be seen, all reference frames arrive at the same result, in accordance with the principle of Galilean relativity.

Problem B.4

Substituting Lorentz transformation into the definition of

$$\eta' = t' - x',$$

we get

$$\eta' = \gamma(t - vx - x + vt) = \gamma(1 + v)(t - x) = k\eta,$$

where

$$k = \sqrt{\frac{1 + v}{1 - v}} = \gamma(1 + v)$$

is the Doppler k-factor. Similarly for τ':

$$\tau' = t' + x' = \gamma(t - vx + x - vt) = \gamma(1 - v)\tau,$$

or

$$\tau' = \tau/k.$$

Note that the product $\eta\tau = \eta'\tau'$ is invariant.

Problem B.5

The rod can not stop all at once. When its front connects with the wall, the rear side will continue moving with the same speed. Some time is required for the "signal" (e.g., the interaction between atoms of a real body) from the front part of the rod to reach the rear part and slow it down. The fastest such a "signal" can propagate is the speed of light.

At the moment when the front of the rod connects with the wall, the measured length of the rod is

$$L = L_0\sqrt{1 - v^2}.$$

At this instant the "signal" from the front starts propagating towards the rear, as shown in Fig. B.42a. The "signal" and the rear of the rod are moving towards each other and "meet" after T seconds. Given the measured length of the rod and the speed of the rear part, we can write

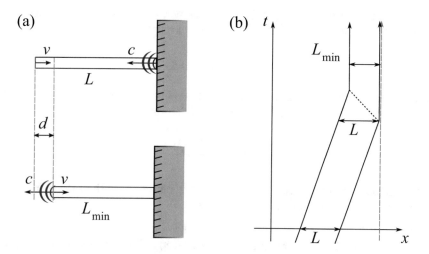

Fig. B.42 (a) Collision of a "rigid" rod with an "absolutely rigid" wall. A signal is sent through the bulk of the rid from the collision point to the left edge. Once the signal reaches the left edge, the latter stops. (b) Spacetime diagram of the collision: The worldlines of the front and the rear of the rod show that $L_{min} < L$

$$L = vT + cT = (1 + v)T,$$

and therefore

$$T = \frac{L}{1 + v}.$$

During this time, the rear of the rod will cover distance $d = vT$, making the rod shorter. Finally, the minimum length of the compressed rod is

$$L_{min} = L - d = L - \frac{vL}{1 + v} = \frac{L}{1 + v}.$$

In terms of the proper length L_0 we can write:

$$L_{min} = \frac{L_0\sqrt{1 - v^2}}{1 + v} = L_0\sqrt{\frac{1 - v}{1 + v}}.$$

Problem B.6

In the reference frame where the ships start their identical engines at the same time, their trajectories will be similar. The distance between them, as measured in this reference frame, will remain constant. The speed of the left ship will be increasing and will be equal to the speed of the right ship. Therefore, the front and the rear of the string will be moving with identical but continuously increasing speeds, all the time maintaining the same measured distance between them. Since the measured length of a moving object is related to the proper length L_0 as

$$L = L_0\sqrt{1 - v^2},$$

the proper length of the string has to be different at the beginning of the motion, compared to any other later moment. In particular, since

$$L_0 = \frac{L}{\sqrt{1 - v^2}},$$

and L remains constant, while v is increasing, the proper length of the string must grow and the string has to snap.

Problem B.7

In this problem the Earth and the Sun are assumed to be inertial reference frames in relative motion. In the reference frame of the Earth two detection events are separated in space and time by the interval

$$\Delta \widehat{s} = (\Delta t, \Delta x) = (0, D),$$

where D is the distance between the LIGO observatories. The components of the same interval measured in the reference frame of the Sun can be found using Lorentz transformation

$$\Delta t' = \gamma(\Delta t - v\Delta x) = -\gamma v D,$$

here v is the orbital speed of the earth. The values for the distance and the speed are given in the regular units: the distance is given in meters, the orbital speed—in meters per second. Using these units, the time interval between the detection events is written as

$$\Delta t' = -\gamma \frac{\bar{v}\bar{D}}{\bar{c}^2}.$$

Using the values

$$\bar{v} = 3 \times 10^4 \text{ m/s},$$

$$\bar{D} = 3 \times 10^6 \text{ m},$$

$$\bar{c} = 3 \times 10^8 \text{ m/s},$$

we get

$$\Delta t' \approx 3 \times 10^{-6} \text{ s}.$$

Fig. B.43 Light signal
bouncing between a pair of
observers, A and B. Using the
Doppler k-factor for the
receding observers and the
radar method, we can find the
times of the reflection events,
as measured by both A and B

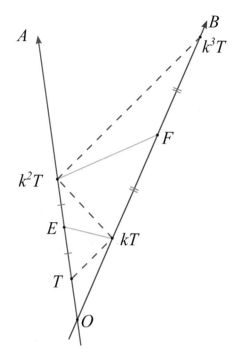

Problem B.8

Figure B.43 shows Minkowski diagram for two observers sending light signals back
and forth. For the relative speed $v = 0.6$, the Doppler k-factor equals 2. Therefore,
the signal emitted at time T by the observer A, will be received and reflected by the
observer B at $2T$ (as measured by B). The reflected signal will be received back by
A at time $4T$ (as measured by A). The signal reflected by A will be finally received
by B at time $8T$.

According to A, the first reflection of the signal by B happens simultaneously
with the event E, which lies half-way between the emission and detection of the
first signal from A:

$$t_E = \frac{T + k^2 T}{2} = T\frac{k^2 + 1}{2} = \frac{5}{2}T.$$

Similarly, the observer B measures the event F to be simultaneous with the first
reflection of the signal by A (at $k^2 T$). It's time, according to B is

$$t_F = \frac{kT + k^3 T}{2} = kT\frac{k^2 + 1}{2} = 5T.$$

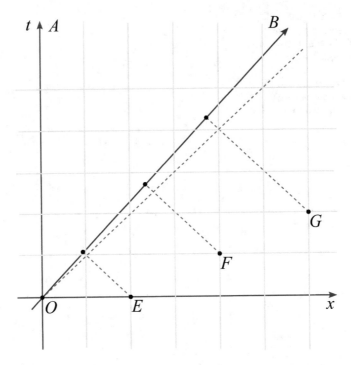

Fig. B.44 Three events are measured and seen by a pair of observers in relative motion. The "visual" order is different from the measured order. Additionally, for the relative speed $v = 0.9$, observers will measure the opposite order of the events E, F, and G

Problem B.9

Figure B.44 shows Minkowski diagram for the reference frame of the observer A. If the observer B is moving relative to A with the speed v, it will measure the time interval between a pair of events

$$\Delta t' = \gamma(\Delta t - v\Delta x).$$

(A) Events E and F will be simultaneous for B if $\Delta t'_{EF} = 0$, which happens for the relative velocity

$$v = \frac{\Delta t_{EF}}{\Delta x_{EF}}.$$

Given that $\Delta t_{EF} = \Delta t_{FG} = 1$s, and $\Delta x_{EF} = \Delta x_{FG} = 2$s, we find that for $v = 0.5$, all three events, E, F, and G will be measured by the observer B to happen at the same time.

(B) For $v = 0.9$ the value of the factor $\gamma = 2.3$. Applying Lorentz transformation to the spacetime coordinates of the events E, F, and G, we find

$$E \to (-4.1, 4.6), \quad F \to (-6.0, 7.1), \quad G \to (-7.8, 9.6).$$

Thus, the measured order of the events is $G - F - E$, opposite to order as measured by the observer A.

(C) From the Minkowski diagram it is clear that the observer B will see the events E, F, and G in the same order as the observer A: $E - F - G$.

The important conclusions from this exercise are: The *measured* order of events can be different from the *apparent* (i.e., visually perceived) for an observer; moreover, the measured order of events can be different for different observers in relative motion. In this example, each pair of events $E - F$, $F - G$, and $E - G$ is connected with a space-like interval, and thus has no absolute order.

Problem B.10

The conservation of energy requires

$$E + \gamma m = E' + m;$$

and the conservation of momentum requires

$$E - \gamma m v = -E',$$

where v is the unknown speed of the electron before the collision. From these equations follows:

$$\gamma = e' - e + 1,$$
$$\gamma v = e' + e.$$

Here we introduced the dimensionless energies

$$e = \frac{E}{m}, \quad e' = \frac{E'}{m}.$$

Using the equality[7]

$$\gamma^2 v^2 = \gamma^2 - 1,$$

---

[7] Easily proved from the definition of γ.

we can write

$$\left(e' + e\right)^2 = \left(e' - e + 1\right)^2 - 1,$$

or, alternatively,

$$\left(e' - e + 1\right)^2 - \left(e' + e\right)^2 = 1.$$

Applying the algebraic identity $a^2 - b^2 = (a - b)(a + b)$, we arrive at

$$\left(2e' + 1\right)\left(1 - 2e\right) = 1,$$

from which follows

$$e' = \frac{e}{1 - 2e} \quad \rightarrow \quad E' = \frac{E}{1 - 2E/m}.$$

The same expression follows directly from the formula (7.19) for the energy of a photon undergoing Compton scattering. Interestingly, the result does not depend on the initial speed of the electron.

Index

© The Author(s), under exclusive license to Springer Nature Switzerland AG 2022
Y. Deshko, *Special Relativity*, Undergraduate Lecture Notes in Physics,
https://doi.org/10.1007/978-3-030-91142-3

Printed in the United States
by Baker & Taylor Publisher Services